# Shape Memory Alloys, Muscle Wires and Robotics

BENJAMIN C. LAWSON

SHAPE MEMORY ALLOYS, MUSCLE WIRES AND ROBOTICS

© 2016 Benjamin C. Lawson

© 2016 Capitol Publishers, LLC
ISBN: 978-1533285607
Printed in the United States of America

To all the people who believe in a better future,
where people behave like people
and robots learn from them.

To a better future besides Anna.

# ABOUT THE AUTHOR

Benjamin C. Lawson was born the 24th May, 1977. In 2001 mastered as a communication engineer and in 2003 as an electronics engineer.

In 2009 he got his Master in International Trade from ISEAD and in 2013, his MBA from IESE Business School, including stays in Madrid, Barcelona, Shanghai and New York.

He has worked in robotics since his very early professional experience. His first project was related to computer vision for forest-fire monitoring. He has also work in autonomous ground vehicle computing and sensing, collaborative robotics and soft robotics.

He added French and German to English and Spanish languages at the age of 14 and, in 2015, Japanese. He has been all his life in the edge of technology and he is the author of three other books related to technology and a book with a study on the meaning and translation of the Tao. He is also the author of more than 50 papers on technology. Lawson is a collaborator of several institutions and associations of technology and he is Senior Member of IEEE since 2012, the biggest engineer association worldwide.

BENJAMIN C. LAWSON

# 1  SHAPE-MEMORY ALLOYS

## 1.1  What is a Shape-Memory Alloy

A shape-memory alloy, so called SMA, is also known as memory metal, smart metal, smart alloy and muscle wire. It is a kind of alloy able to be deformed in environmental temperature and then return to its pre-deformed shape when heated.

Shape-memory alloys have applications in robotics and automotive, aerospace and biomedical industries. They can be highly useful when weight, power consumption or reduced space are critical design requirements, making them better than traditional hydraulic, pneumatic or motor actuators.

The discovery of the shape-memory effect took place in the 1930s. A. Ölander discovered the pseudo-elastic behavior of the Au-Cd alloy in 1932. Greninger and Mooradian observed the deformation and shape recovering phenomenon by decreasing and increasing the temperature of a Cu-Zn alloy in 1938. This basic phenomenon was widely reported a decade later by Kurdjumov and Khandros, in 1949, and also by Chang and Read, in 1951.

The nickel-titanium alloys were first developed between 1962 and 1963 by the United States Naval Ordnance Laboratory and commercialized under the trade name Nitinol (an acronym for Nickel Titanium Naval Ordnance Laboratories).

There is another type of SMA, called a ferromagnetic shape-memory alloy (FSMA), that changes shape under strong magnetic fields. These materials are of particular interest as the magnetic response tends to be faster and more efficient than temperature-induced responses, but they are usually out of the scope of robotics, so we won't conver them in this book.

Metal alloys are not the only thermally-responsive materials; shape-memory polymers have also been developed, and became commercially available in the late 1990s. However, they change their shape much slower than metal alloys (in a period of some seconds, typically) so they are useless for the majority of robotics applications and won't be covered in this book either.

A wide variety of alloys exhibit the shape-memory effect. The transformation temperature can be changed by adjusting alloying constituent proportions. Some experimented alloys are shown in Table 1.

| Alloy Symbol | Alloy Constituents | Proportion Range |
|---|---|---|
| Ag-Cd | Silver, cadmium | 44/49% Cd |
| Au-Cd | Gold, cadmium | 46.5/50% Cd |
| Cu-Al-Ni | Copper, aluminum, nickel | 14/14.5% Al and 3/4.5% Ni |
| Cu-Sn | Copper, tin | 15% Sn |
| Cu-Zn | Copper, zinc | 38.5/41.5% Zn |
| Cu-Zn-Si | Copper, zinc, silicon | |
| Cu-Zn-Al | Copper, zinc, aluminum | |
| Cu-Zn-Sn | Copper, zinc, tin | |
| Fe-Pt | Iron, platinum | 25% Pt |
| Mn-Cu | Manganese, copper | 5/35% Cu |
| Fe-Mn-Si | Iron, manganese, silicon | |
| Co-Ni-Al | Cobalt, nickel, aluminum | |
| Co-Ni-Ga | Cobalt, nickel, gallium | |
| Ni-Fe-Ga | Nickel, iron, gallium | |
| Ti-Nb | Titanium, niobium | |
| Ni-Ti | Nickel, titanium | 55-60% Ni |
| Ni-Ti-Hf | Nickel, titanium, hafnium | |
| Ni-Ti-Pd | Nickel, titanium, palladium | |
| Ni-Mn-Ga | Nickel, manganese, gallium | |

*Table 1. Some known shape-memory alloys*

The two main types of shape-memory alloys are copper-aluminum-nickel, and nickel-titanium alloys, commonly called "nitinol" (the same as the brand but without the initial capital letter) but SMAs can also be created by alloying zinc, copper, gold and iron. Although iron-based and copper-based SMAs, such as Fe-Mn-Si, Cu-Zn-Al and Cu-Al-Ni, are commercially available and cheaper than nitinol, the latter is preferable for most applications due to its stability, practicability and superior thermo-mechanic performance.

# 1.2 Crystal Structures of SMAs

SMAs can exist in two different phases, with three different crystal structures, described below, and six possible transformations:

- **Twinned martensite**: martensite is the low temperature phase and it is represented by 'M'. Twinned martensite is the **cold form without deformation**, so represented by 'M$^t$' and, when the alloy is in this form, each crystal is in its "self-accomodated" orientation direction (so called variant). When the material is heated from the martensitic phase, the crystal structure transforms to austenite. This transition is called *reverse transformation*, during which there is no associated shape change if the material has remained in twinned martensite.

- **Detwinned martensite**: is represented by 'M$^d$'. In this form, most of the crystals in the alloy have the same orientation direction. It is the **cold form with deformation**. The phenomenon of deforming the alloy is called the *martensitic transformation*. It occurs when applying a mechanical load to the material in the twinned martensitic phase (at low temperature). The detwinning crystal process results in a macroscopic shape change, where deformed configuration is retained when the load is released. However, when the material is heated above a certain temperature, a reverse phase transformation will occur (from detwinned martensite to austenite) and it will lead to complete original shape recovery.

- **Austenite**: is the **high temperature** phase and it is represented by 'A'. Upon cooling in the absence of an applied load, the crystal structure changes from austenite to martensite. The phase transition from austenite to martensite is termed the *forward transformation*.

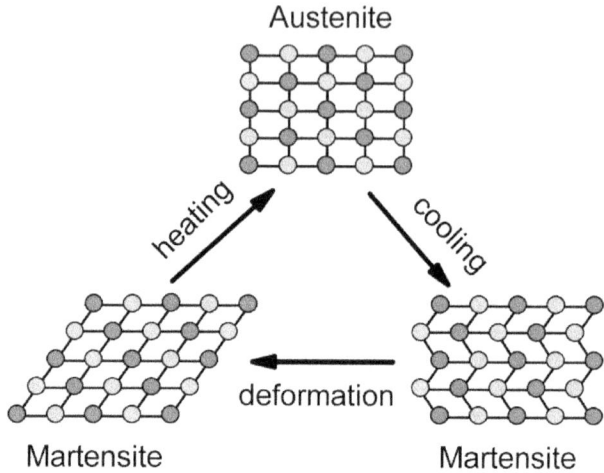

*Fig. 1. Crystal structures of SMAs*

There are four characteristic temperatures associated with the phase transformation. During the forward transformation, austenite, under zero load, begins to transform to twinned martensite at the *martensitic start temperature* ($M_s$) and completes transformation to martensite at the *martensitic finish temperature* ($M_f$). At this stage, the transformation is complete and the material is fully in the twinned martensitic phase.

Similarly, during heating, the reverse transformation initiates at the *austenitic start temperature* ($A_s$) and the transformation is completed at the *austenitic finish temperature* ($A_f$).

The transition from the martensite phase to the austenite phase is only dependent on temperature and stress, not time. Therefore, when engineering an SMA actuator, it is needed to know the desired temperature, usually got through the Joule effect by using a current through the SMA wire, so a precise closed-loop control system is frequently used to actuate the device.

The alternate use of the two martensite and austenite phases of an SMA is what produces movement in robotics. So actuators are usually based on wires made from SMAs, especially a component called nitinol. However, the repeated use of the shape-memory effect may lead to a shift of the characteristic transformation temperatures. This effect is known as functional fatigue, as it is closely related with a change of microstructural and functional properties of the material.

There is also a maximum temperature at which SMAs can no longer be stress induced is called $M_d$, where the SMA is permanently deformed. When engineering the actuators, it is used to set the initial actuator form.

It is also very important to know all the SMAs have hysteresis when changing from austenite to martensite and vice-versa. It is shown in Fig. 2. $\xi(T)$ represent the martensitic transformation fraction (0 meaning no deformation and 1 representing full deformation). The difference between the heating transition and the cooling transition gives rise to hysteresis where some of the mechanical energy is lost in the process. The shape of the curve depends on the material properties of the specific SMA used.

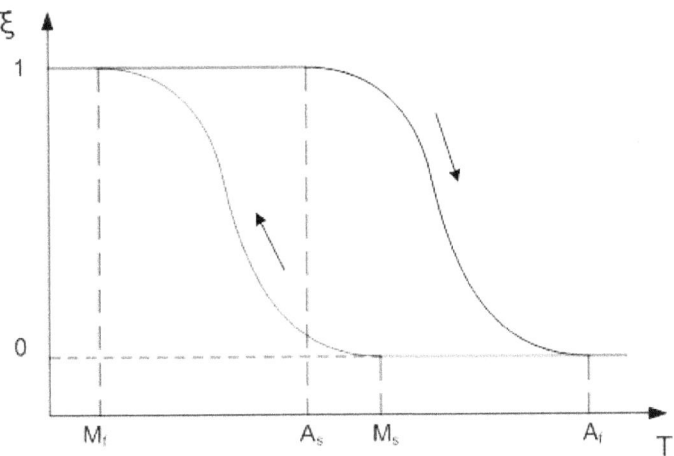

*Fig. 2. Hysteresis curve when heating (black) and cooling & loading (red) an SMA*

In Table 2 are shown the four temperatures for the nickel, titanium-based SMAs. Although original nitinol temperature for finishing austenite phase is almost 90°C, the alloy typically used for robotics has lower temperatures, and austenite is finished at 70°C.

| NiTi-based SMA | $M_f$ | $M_s$ | $A_s$ | $A_f$ |
|---|---|---|---|---|
| $Ti_{50}Ni_{50}$ (original nitinol) | 15 | 55 | 80 | 89 |
| $Ti_{49.5}Ni_{50.5}$ | -78 | -19 | 9 | 53 |
| $Ti_{49}Ni_{51}$ | -153 | -114 | -89 | -40 |
| $Ti_{49}Ni_{41}Cu_{10}$ | 8 | 30 | 35 | 50 |
| $Ti_{50}Ni_{40}Cu_{10}$ | 21 | 41 | 53 | 67 |
| $Ti_{44}Ni_{47}Nb_9$ | -175 | -90 | -85 | -35 |
| $Ti_{42.2}Ni_{49.8}Hf_8$ | 50 | 69 | 111 | 142 |
| $Ti_{40.7}Ni_{49.8}Hf_{9.5}$ | 61 | 90 | 118 | 159 |
| $Ti_{40.2}Ni_{49.8}Hf_{10}$ | 103 | 128 | 182 | 198 |
| $Ti_{35.2}Ni_{49.8}Hf_{15}$ | 95 | 136 | 140 | 210 |
| $Ti_{30.2}Ni_{49.8}Hf_{20}$ | 127 | 174 | 200 | 276 |
| $Ti_{48}Ni_{47}Zr_5$ | 20 | 65 | 75 | 138 |
| $Ti_{43}Ni_{47}Zr_{10}$ | 45 | 100 | 113 | 165 |
| $Ti_{38}Ni_{47}Zr_{15}$ | 100 | 175 | 175 | 230 |
| $Ti_{33}Ni_{47}Zr_{20}$ | 205 | 275 | 265 | 330 |
| $Ti_{50}Pd_{50}$ | 550 | 563 | 580 | 591 |
| $Ti_{50}Ni_{20}Pd_{30}$ | 208 | 241 | 230 | 241 |
| $Ti_{50}Ni_{10}Pd_{40}$ | 387 | 403 | 419 | 427 |
| $Ti_{50}Ni_5Pd_{45}$ | 467 | 486 | 503 | 509 |
| $Ti_{50}Ni_{45}Pt_5$ | 10 | 29 | 36 | 49 |
| $Ti_{50}Ni_{40}Pt_{10}$ | -8 | 18 | -27 | 36 |
| $Ti_{50}Ni_{30}Pt_{20}$ | 241 | 300 | 263 | 300 |
| $Ti_{50}Ni_{20}Pt_{30}$ | 537 | 619 | 626 | 702 |

*Table 2. The four temperatures (°C) for most frequent NiTi SMAs*

# 1.3 One-Way Vs. Two-Way Shape Memory

Shape-memory alloys have different shape-memory effects. Two common effects are one-way and two-way shape memory. The procedures are very similar: starting from martensite (a), adding a reversible deformation for the one-way effect or severe deformation with an irreversible amount for the two-way (b), heating the sample (c) and cooling it again (d).

## 1.3.1 One-Way Memory Effect

When a shape-memory alloy is in its cold state (below $A_s$), the metal can be bent or stretched and will hold those shapes until heated above the transition temperature. Upon heating, the shape changes to its original. When the metal cools again it will remain in the hot shape, until deformed again.

With the one-way effect, cooling from high temperatures does not cause a macroscopic shape change. A deformation is necessary to create the low-temperature shape. On heating, transformation starts at $A_s$ and is completed at $A_f$ (typically 2 to 20 °C or hotter, depending on the alloy or the loading conditions). $A_s$ is determined by the alloy type and composition and can vary between −150 °C and 200 °C.

This is the effect normally used in robotics, where the deformation is performed by the use of a spring or using a couple of SMA wires pulling one against each other.

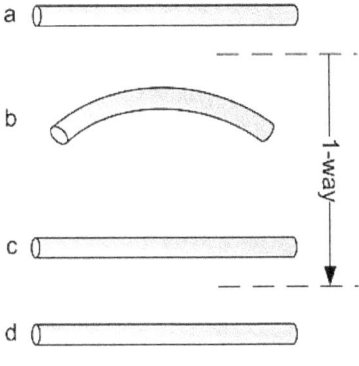

*Fig. 3. 1-way shape memory*

## 1.3.2   Two-Way Memory Effect

The two-way shape-memory effect is the effect that the material remembers two different shapes: one at low temperatures, and one at the high-temperature shape. A material that shows a shape-memory effect during both heating and cooling is said to have two-way shape memory. This can also be obtained without the application of an external force (intrinsic two-way effect). The reason the material behaves so differently in these situations lies in training. Training implies that a shape memory can "learn" to behave in a certain way. Under normal circumstances, a shape-memory alloy "remembers" its low-temperature shape, but upon heating to recover the high-temperature shape, immediately "forgets" the low-temperature shape. However, it can be "trained" to "remember" to leave some reminders of the deformed low-temperature condition in the high-temperature phases. A shaped, trained object heated beyond a certain point will lose the two-way memory effect.

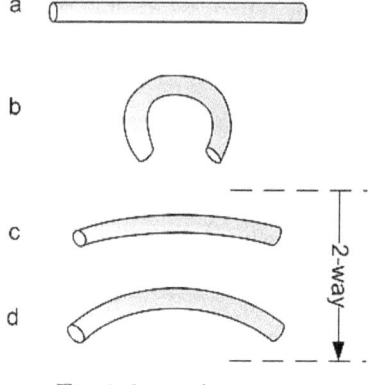

*Fig. 4. 2-way shape memory*

## 1.4 Super-Elasticity

SMAs also display super-elasticity, which is characterized by recovery of unusually large strains. Instead of transforming between the martensite and austenite phases in response to temperature, this phase transformation can be induced in response to mechanical stress. When SMAs are loaded in the austenite phase, the material will transform to the martensite phase above a critical stress, proportional to the transformation temperatures. Upon continued loading, the twinned martensite will begin to detwin, allowing the material to undergo large deformations. Once the stress is released, the martensite transforms back to austenite, and the material recovers its original shape. As a result, these materials can reversibly deform to very high strains –up to 8 percent.

## 1.5 How SMAs Are Produced

Shape-memory alloys are typically made by casting, using vacuum arc melting or induction melting. These are specialist techniques used to keep impurities in the alloy to a minimum and ensure the metals are well mixed. The ingot is then hot rolled into longer sections and then drawn to turn it into wire.

The way in which the alloys are "trained" depends on the properties wanted. The "training" dictates the shape that the alloy will remember when it is heated. This occurs by heating the alloy so that the dislocations re-order into stable positions, but not so hot that the material recrystallizes. They are heated to between 400 °C and 500 °C for 30 minutes, shaped while hot, and then are cooled rapidly by quenching in water or by cooling with air.

# 1.6 Properties for Engineering

The copper-based and nitinol shape-memory alloys are considered to be engineering materials. These compositions can be manufactured to almost any shape and size.

The yield strength of shape-memory alloys is lower than that of conventional steel, but some compositions have a higher yield strength than plastic or aluminum. The yield stress for nitinol can reach 500 MPa. The high cost of the metal itself and the processing requirements make it difficult and expensive to implement SMAs into a design. As a result, these materials are used in applications where the super elastic properties or the shape-memory effect can be exploited. The most common application is in actuation.

One of the advantages to using shape-memory alloys is the high level of recoverable plastic strain that can be induced. The maximum recoverable strain these materials can hold without permanent damage is up to 8% for some alloys, although usually manufacturers do not recommend a strain over 5%.

# 1.7 Applications of SMAs

## 1.7.1 Automotive

The first high-volume SMA product is an automotive valve used to control low pressure pneumatic bladders in a car seat that adjust the contour of the lumbar bolsters. More than 5 million actuators are sold every year. The overall benefits of SMA over traditionally-used solenoids in this application are a lower noise, electromagnetic compatibility, less weight, a smaller form factor and reduced power consumption.

Other uses of SMA in automobiles are the replacement of heavier motorized actuators to open and close the hatch vent that releases air from the trunk, making it easier to close, electric generators to generate electricity from exhaust heat and on-demand air dams to optimize aerodynamics at various speeds.

## 1.7.2 Smartphones

The second high volume application of SMA is an autofocus (AF) actuator for a smartphone. Research is also being made to build an optical image stabilisation (OIS) module driven by SMA wires.

## 1.7.3 Medicine

Shape-memory alloys are frequently applied in some medical areas. Specific uses are as fixation devices for osteotomies in orthopedic surgery, in dental braces to exert constant tooth-moving forces on the teeth and as tools for surgery.

Nitinol has the important advantage of being compatible with the human body, so it does not produce rejection. Since it was introduced in the late 1980s, it has been used to build micro-pumps for helping heart or kidneys malfunction and as surgery tool to clean blocked arteries.

Thus, nitinol is considered for minimally invasive endovascular medical applications. Although it is more expensive than stainless steel, its self-expanding properties when manufactured to BTR (Body Temperature Response), have provided an attractive alternative to traditional methods in some typical uses:

- **Stents**: a cardiovascular stent is a device with an approximately tubular shape, inserted in coronary arteries when they are occluded. When put in place, it allows the artery remain open and the blood can pass by. Approximately 50% of all peripheral vascular stents currently provided on the worldwide market are manufactured with nitinol

- **Bypass**: when the occlusion continues after a stent has been put in place, bypass is the recommended solution. In this case, the surgery consists of making a cut in the artery where healthy and grafting it to the coronary vessel, avoiding the blood circulation through the affected artery fragment. Nitinol is being investigated to be used as a joint between arteries. In this case, a nitinol ring is inserted between two arteries ends and, when it is heated by the body's temperature, it expands and fixes boths ends. After that, the surgeon sutures the ring to the arteries for permanent fastening.

- **Orthopedic surgery:** memory metal has been utilized in orthopedic surgery as a fixation-compression device for osteotomies, typically for lower extremity procedures. The device, usually in the form of a large staple, is stored in a refrigerator in its malleable form and is implanted into pre-drilled holes in the bone across an osteotomy. As the staple warms it returns to its non-malleable state and compresses the bony surfaces together to promote bone union.

- **Orthodontics:** a major area of development for nitinol is dentistry. One example is the prevalence of dental braces using SMA technology to exert constant tooth-moving forces on the teeth. The first nitinol arch wire was developed in 1976 by Dr. George Andreasen, after Dr. William J. Buehler first started to experiment with nitinol in the Naval Ordnance Laboratory of the United States in 1958. After that, in 1986, Dr. Fujio Miura developed sentalloy, the first super-elastic SMA, made of nickel and titanium. Super-elasticity is a great advantage for orthodontics: at low temperature, sentalloy wires are flexible and easy to manipulate. Shape memory gets activated at exactly 37.25°C (99.5°F). Then the wire returns to it original arch shape with a soft but constant force, producing the teeth ideal, physiological movement with minimum damage to the wire. Therefore, there is no need to change the sentalloy wires frequently.

## 1.7.4 Aircraft and spacecraft

Boeing, General Electric Aircraft Engines, Goodrich Corporation, NASA, Texas A&M University and All Nippon Airways developed the Variable Geometry Chevron using nitinol. Such a variable area fan nozzle

(VAFN) design will allow for quieter and more efficient jet engines. In 2005 and 2006, Boeing conducted successful flight testing of this technology.

SMAs are being explored as vibration dampers for launch vehicles and commercial jet engines. The large amount of hysteresis observed during the super-elastic effect allow SMAs to dissipate energy and dampen vibrations. These materials show promise for reducing the high vibration loads on payloads during launch as well as on fan blades in commercial jet engines, allowing for more lightweight and efficient designs. SMAs also exhibit potential for other high shock applications such as ball bearings and landing gear.

## 1.7.5 Robotics

There have been some studies on using these materials in robotics. The most used form of SMAs in robotics is the "muscle wire", consisting of nitinol wires of different diameters, used to actuate the movements of a robot. The pros of muscle wire are a smaller form factor, high resistance, high life without maintenance and low power consumption. However, there are some cons, as slow response times and large hysteresis.

The most famous laboratory robot made with nitinol is the hobbyist robot Stiquito, with the form of an insect using muscle wires to move the legs (see Fig. 5). A prosthetic hand that can almost replicate the motions of a human hand was introduced in 2005. Other biomimetic applications are also being explored.

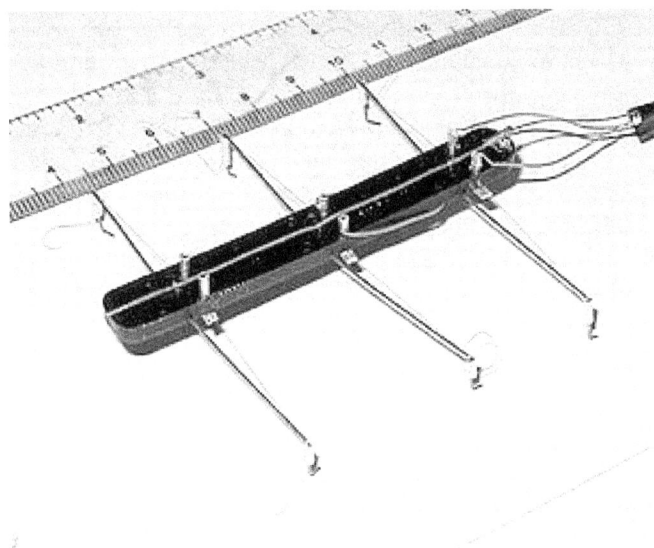

*Fig. 5. Stiquito is probably the most famous muscle wire robot*

## 1.7.6   Other uses

There are many other uses for SMAs in the industry. The following are less frequent uses, but research and increasing knowledge about SMAs are growing the situations where SMAs are found better than traditional devices:

- **Civil Structures:** SMAs find a variety of applications in civil structures such as bridges and buildings. One such application is Intelligent Reinforced Concrete (IRC), which incorporates SMA wires embedded within the concrete. These wires can sense cracks and contract to heal macro-sized cracks. Another application is active tuning of structural natural frequency using SMA wires to dampen vibrations.

- **Engines:** experimental solid state heat engines, operating from the relatively small temperature differences in cold and hot water reservoirs, have been developed since the 1970s.

- **Piping:** the first consumer commercial application was a shape-memory coupling for piping, e.g. oil line pipes for industrial applications, water pipes and similar types of piping for consumer/commercial applications.

- **Glasses:** the are some commercial eyeglass frames made from titanium-containing SMAs that have their transition temperature set below the expected room temperature. This allows the frames to undergo large deformation under stress, yet regain their intended shape once the metal is unloaded again. The very large apparently elastic strains are due to the stress-induced martensitic effect, where the crystal structure can transform under loading, allowing the shape to change temporarily under load. This means that eyeglasses made of shape-memory alloys are more robust against being accidentally damaged.

# 2 SHAPE-MEMORY ALLOYS IN ROBOTICS

With its unique ability to contract on demand, shape-memory alloys present many intriguing possibilities for robotics as actuators. An actuator wire is usually made from nitinol, and marketed under a certain number of brand names including Nitinol, Flexinol, BioMetal, and Muscle Wire. As we have already seen, nitinol actuator wires are able to contract with significant force and can be better than servo motors, pneumatic actuators and solenoids in some specific environments.

Flexinol and Muscle Wire are both trademarks of the American company Dynalloy. Flexinol is referenced in many books and research articles and has been specified in a number of student and hobby-level robotics projects. Perhaps most famously, Flexinol was used to operate a dust sensor installed on the Sojourner rover which landed on Mars as part of the Mars Pathfinder mission July 4, 1997 (see Fig. 6).

*Fig. 6. Flexinol was used to operate a dust collector on the Sojourner rover, landed on Mars July 4, 1997.*

Although nitinol actuator wires will only contract by about 3-7% of its length in the transition, it is accompanied by such force that it can easily be applied through levers and linkages to achieve much larger movements.

## 2.1 Dependence Between Load and Temperature

The four phase-transition temperatures of SMAs depend on how much stress is applied by the mechanical load responsible for the desired deformation.

So, the load applied must be sufficiently large to start the detwinning process. The minimum stress required for detwinning initiation is termed the *detwinning start stress* ($\sigma_s$). Sufficiently high-load levels will result in complete detwinning of martensite, where the corresponding stress level is called the *detwinning finish stress* ($\sigma_f$).

When the material is cooled with a mechanical load greater than $\sigma_s$

applied in the austenitic phase, the phase transformation will result in the direct formation of detwinned martensite, producing a shape change. Reheating the material will result in shape recovery while the load is still applied. A schematic of the above-described loading path is shown in Fig. 7. Recognizing that the forward and reverse transformations occur over a range of temperatures ($M_s$ to $M_f$, $A_s$ to $A_f$) for a given alloy composition, we can construct transformation regions in the stress-temperature space. The transformation temperatures strongly depend on the magnitude of the applied load, with higher values of applied load leading to higher transformation temperatures.

As a consequence, the transformation regions representing the $A \rightarrow M^d$ and $M^d \rightarrow A$ transformations have a positive slope in stress-temperature space.

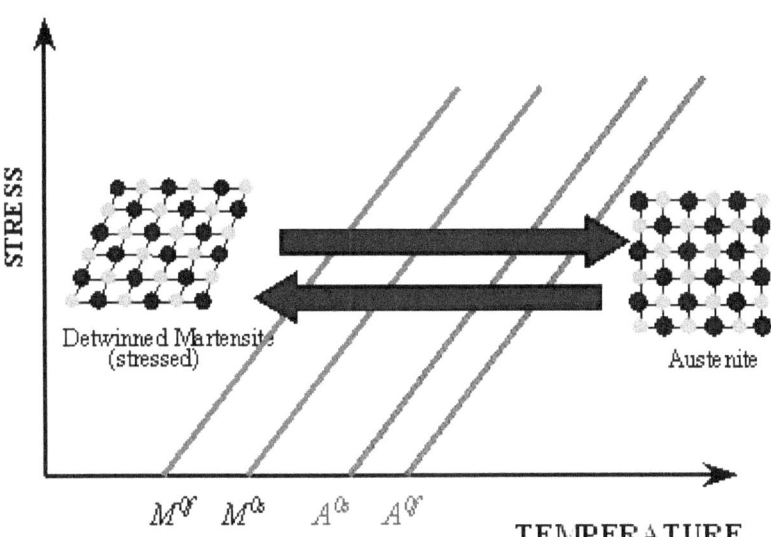

*Fig. 7. Temperature-induced phase transformation in the presence of applied load*

When speaking of robotics applications, the properties of an actuator wire are highly dependent on the manufacture of the nitinol and the preparation of the wire. As seen, nitinol can be formulated to transition across a wide range of temperatures. When intended to be an actuator, nitinol is usually formulated to transition at a temperature achievable with electric current. As an example, Flexinol is available in a both low temperature (70 °C) and high temperature (90 °C) formulations to support different applications. However, remember that unprepared nitinol is not an actuator; when heat is applied the only result will be that it gets hotter.

On the other side, the actuator wire diameter will dictate the maximum force that can be applied. Diameter also dictates the wire's resistance to electric current, the available current required to reach transition and the time the wire will require to heat and cool.

## 2.2 Electronic Control

Nitinol resistance depends on the diameter of the wire. As an example, the widespread 0.006" (0.15 mm) wire presents a resistance of 0.5 $\Omega$/cm.

Success can be checked experimentally as well. If the wire begins to contract immediately after current is removed, then it is probably at about the correct temperature. If on the other hand, the wire does not begin to contract immediately after current is removed, then it is probably running too hot and in danger of being damaged.

## 2.3 Pressure Regulation

Nitinol usually requires a minimum pressure of 10,000 to 20,000 PSI to be deformed. On the other side, the pressure produced in the reverse transformation, when the material is heated, is much higher, between 35,000 and 70,000 PSI.

## 2.4 Training the Structure

To give nitinol the desired shape, it has to be heated at a much higher temperature. Available commercial products usually need above 540 °C to set the new crystal structure to be remembered in austenite stage. However, temperature should never go above fusion point, where the crystal structure should be destroyed and the material would loss its shape-memory property. It is normally around 1,250 °C (2,282 °F).

## 2.5 Benefits of Nitinol Wires

The potential advantages to working with nitinol actuators are significant. Solutions can be physically smaller when compared to a servo or solenoid. Weight considerations make often a difference. In Fig. 8 a comparison with other technologies in term of power and weight ratio is provided.

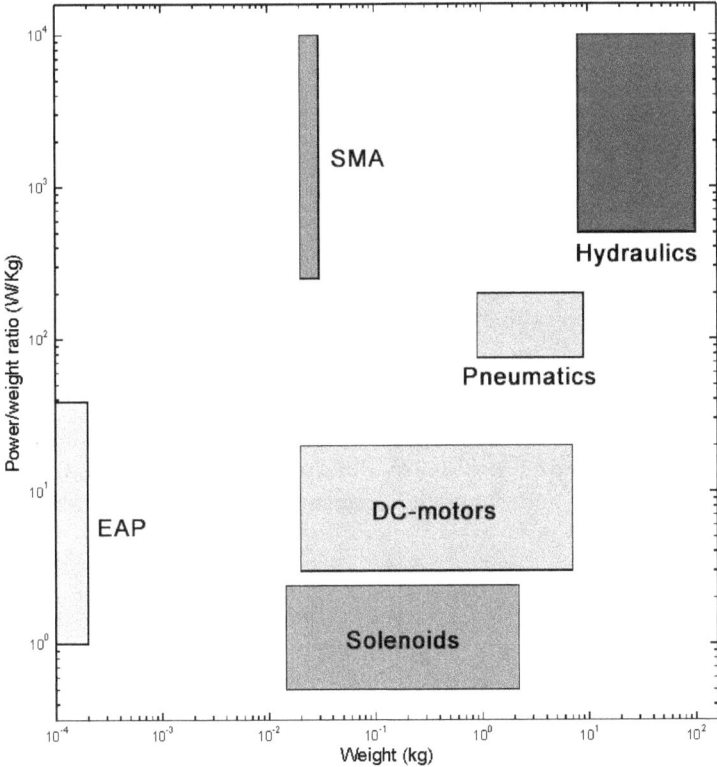

*Fig. 8. Comparison of different technologies for robotics actuators*

Additionally, because the activation mechanism is heat instead of electromagnetism, these solutions do not introduce magnetic noise to a circuit, so they become interesting when electromagnetic compatibility is a must.

**Strength**

Nitinol wires have the ability to pull relative high weight. It makes them highly useful when space for actuators is a problem.

**Quiet movement**

Nitinol wires do not produce noise when actuating, which is highly appreciated for humanoid and domestic robots, for instance.

# 2.6 Practical Limitations

Nitinol wires have many advantages over traditional actuators, but do suffer from a series of limitations that may impede practical application. Designing with nitinol brings its own set of challenges. Working with the thin, tough wires can be physically difficult and a substantial current can be required to achieve transition.

Nevertheless, perhaps the greatest challenge lies in the fact that actuator wire must be stretched back into its lengthened state with each activation cycle. This stretching requires that a counter-balancing force be engineered into the project.

Once all design considerations have been taken into account, there are some other physical and structural limitations for shape-memory alloy actuators, as shown below.

## 2.6.1    Response Time and Asymmetry

SMA actuators are typically actuated electrically, where an electric current results in Joule heating. Deactivation typically occurs by free convective heat transfer to the ambient environment. Consequently, SMA actuation is typically asymmetric, with a relatively fast actuation time and a slow recovery time. Typically, thin nitinol wires (50 μm diameter) can get contracted in around 0.01s, but will take around 0.1s to recover. Thicker wires can take up to 3s to recover, making them useless for most robotics applications. As a rule of thumb, you can estimate 3X contraction time for recovering.

A number of methods have been proposed to reduce SMA deactivation time, including forced convection, and lagging the SMA with a conductive material in order to manipulate the heat transfer rate.

Novel methods to enhance the feasibility of SMA actuators include the use of a conductive "lagging". this method uses a thermal paste to rapidly transfer heat from the SMA by conduction. This heat is then more readily transferred to the environment by convection as the outer radii (and heat transfer area) is significantly greater than for the bare wire. This method results in a significant reduction in deactivation time and a symmetric activation profile. As a consequence of the increased heat transfer rate, the required current to achieve a given actuation force is increased.

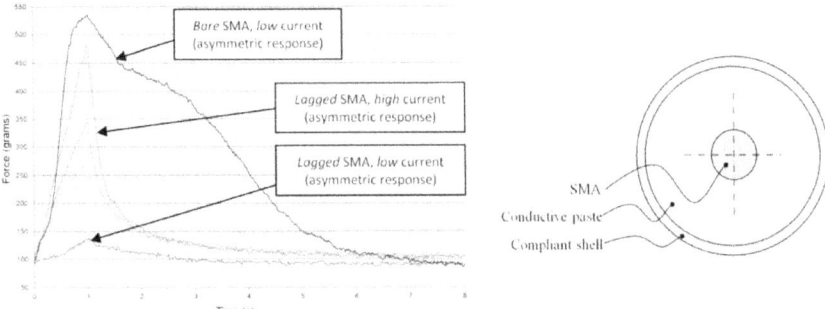

*Fig. 9. Comparative force-time response of bare and lagged nitinol*

## 2.6.2   Structural and Functional Fatigue

SMA is subject to structural fatigue –a failure mode by which cyclic loading results in the initiation and propagation of a crack that eventually results in catastrophic loss of function by fracture. The physics behind this fatigue mode is accumulation of microstructural damage during cyclic loading. This failure mode is observed in most engineering materials, not just SMAs.

SMAs are also subject to functional fatigue, a failure mode not typical of most engineering materials, whereby the SMA does not fail structurally but loses its shape-memory or super-elastic characteristics over time. As a result of cyclic loading (both mechanical and thermal), the material loses its ability to undergo a reversible phase transformation. For example, the working displacement in an actuator decreases with increasing cycle numbers. The physics behind this is gradual change in microstructure —more specifically, the buildup of accommodation slip dislocations. This is often accompanied by a significant change in transformation temperatures.

### 2.6.3 Unintended Actuation

SMA actuators are typically actuated electrically by Joule heating. If the SMA is used in an environment where the ambient temperature is uncontrolled, unintentional actuation by ambient heating may occur.

## 2.7 Flexinol Wire

Flexinol is a trademark of Dynalloy Inc. It is a nitinol wire, trained to be used as an actuator in robotics. While the transformation temperature of nitinol alloys can be adjusted from over 100 °C (212 °F) down to cryogenic temperatures, the temperature for the Flexinol actuator wire has been chosen to be between 60 and 110 °C (between 140 and 230 °F). This allows easy heating with modest electrical currents applied directly through the wire, and quick cooling to below the transformation temperature as soon as the current is stopped.

When used in recommended ranges of temperature and load, it can last 1,000,000 cycles. That is much longer than the lifetime for typical nitinol wires, around only 1,000 cycles.

Although the wire can be contracted up to 8% to 10%, it is recommended not to exceed 6% to preserve its million-cycle life.

Main properties for flexinol wires are in Table 3. Please advise there are two types of wires for each diameter, LT and HT. HT wires have a higher transition temperature (90 °C instead of 70 °C), that means an about 50% shorter cooling period than LT equivalents, under normal conditions (quiet air). If air convection or fluid immersion is used, cooling period can be reduced up to 10X.

The figures indicated for current to contract in a second are only approximate since room temperatures, air currents and heat sinking of specific devices vary. On small diameter wires (up to 150 μm) current that heats the wire in a second can typically be left on without overheating it.

For the cooling times in the table, the last 0.5% of deformation has not been considered.

| Diameter ("/μm) | Resistance (Ω/m) | Pull force (g) | Cooling deformation force (g) | Approx. current to contract in a second (mA) | Cooling time LT/HT (s) |
|---|---|---|---|---|---|
| 0.001/25 | 1425 | 8.9 | 3.6 | 45 | 0.18/0.15 |
| 0.0015/38 | 890 | 20 | 8 | 55 | 0.24/0.2 |
| 0.002/50 | 500 | 36 | 14 | 85 | 0.4/0.3 |
| 0.003/76 | 232 | 80 | 32 | 150 | 0.8/0.7 |
| 0.004/100 | 126 | 143 | 57 | 200 | 1.1/0.9 |
| 0.005/130 | 75 | 223 | 89 | 320 | 1.6/1.4 |
| 0.006/150 | 55 | 321 | 128 | 410 | 2/1.7 |
| 0.008/200 | 29 | 570 | 228 | 660 | 3.2/2.7 |
| 0.01/250 | 18.5 | 891 | 356 | 1050 | 5.4/4.5 |
| 0.012/310 | 12.2 | 1280 | 512 | 1500 | 8.1/6.8 |
| 0.015/380 | 8.3 | 2004 | 802 | 2250 | 10.5/8.8 |
| 0.02/510 | 4.3 | 3560 | 1424 | 4000 | 16.8/14 |

*Table 3. Flexinol specifications*

# 3 BUILDING A NITINOL ACTUATOR

## 3.1 Atachment of Nitinol Wires

To be used as an actuator, Flexinol must be attached both to the mechanical elements where it will cause movement, as well as to the electric circuit that will energize it. Crimping the wire is the only viable option. The thickness of the wire changes slightly as it goes through transition and therefore solder or glue solutions will ultimately fail. Although it is possible to buy Flexinol pre-crimped, in most cases design considerations force to crimp the wires on customized wire lengths. The hair-thin nature of most Flexinol gauges makes crimping difficult, but it is certainly possible to get a good crimp with hand tools. There are many potential approaches. In the original Stiquito plans, Mills ties a small, loose knot in the wire and crimps with a section of 1/16 inch aluminum tubing. Roger Gilbertson, author of the Muscle Wires Project Book, recommends the N-scale rail joiners which are used to connect model railroad conductive track. Many builders will use standard solderless terminals from other electronics applications. A good and simple solution for a temporary crimp that can be repositioned is to use standard machine screws and hex nuts.

However, to physically fix the nitinol wire it is recommended to use wire terminal by pressure, as those shown in Fig. 10. These terminals have the advantage of preserving the wire fully straight and allowing to trim exact lenghts of wires as desired.

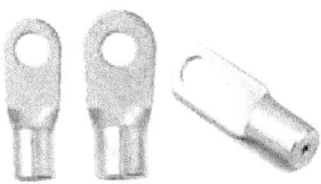

*Fig. 10. Recommended wire terminals for nitinol fixation*

## 3.2 Electrical Stimulus

In most applications, electricity is used to generate the internal heat needed for transition. Proper control of the current is essential for satisfactory results. Too little current and the wire will fail to contract. Too much current and the wire will overheat, becoming stressed and losing its shape memory properties. In between too little and too much, variations in supplied current will affect the heating and cooling times. Some guidelines as to how much current is required to cause a Flexinol wire of a given diameter to contract in one second is provided in Table 3, as well as the approximate electrical resistance. Note that Flexinol has fairly high resistance compared with copper wire and other common conductors. In some applications it is possible to activate the wire without additional resistance in the circuit.

## 3.3 Bias Force

Al we have already seen, Flexinol does not automatically return to its lengthened state when cooled. A force must be applied to the wire to pull it back to length. This opposing force, used to stretch the wire, is called the bias force.

In some applications it is possible to use the natural motion of the mechanism itself to provide the recovery force. For example, in the case of an arm that lifts a dead weight, when power is removed and the Flexinol cools, the weight of the payload will stretch the wire back to its extended length. Likewise, a robot leg that lifts the machine itself can use the weight of the robot to refresh the Flexinol.

When the restoration through intrinsic actuator weigth is not possible, the usual mechanism to restore the long shape of a nitinol wire is by applying a constant force provided either by a spring or a static weight.

The speed and force provided by a nitinol wire contraction depend on how fast its temperature increases and the maximum temperature achieved. A millisecond-time reaction can be provided through high current pulses, but in this case it is crucial to calculate the mass and speed to be moved. If the mass to be moved present higher inertia than the wire maximum traction resistance, the latter will get broken.

The phase transformation, or crystal change, that causes the memory effect has more driving force than the strength of the parent material, so one must use care not to exceed that yield strength. The yield strength of actuator high temperature phase is over 345 Mpa (50,000 PSI) and, on a single pull, the wire can exert this force. To have repeat cycling, however, one should use no more than 2/3 this level and forces of 138 Mpa (20,000 PSI) or below give the best repeat cycling with minimal permanent deformation of the wire.

Additionally, when designing the restoring mechanism, one might call for a load as close to the minimum required refresh force as possible to provide most work energy from the actuator, or one might choose a more weighted load in order to get a quicker restoration speed, at the expense of less force available for contraction.

Anyway, while a normal bias spring increases its force as the actuator contracts, reverse bias forces can be used to decrease the opposite force as the actuator wire contracts. This bias type allows an increase in the actuator stroke when contracting, from a typical 3-4% value to a flex up to 7%. A common example of reverse bias force is the one provided by a magnet. Reverse bias forces have the difficulty to be precisely calculated in the design and, when built, a calibration range has to be provided in order to get the best from the device.

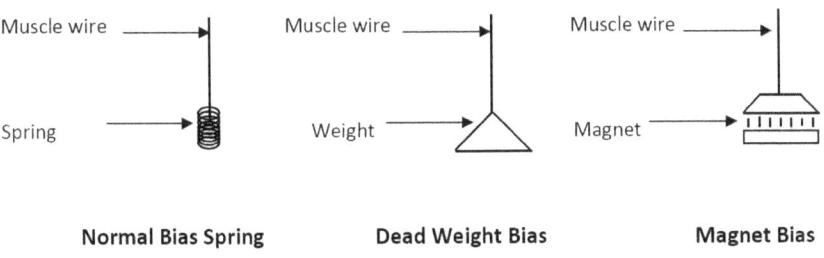

**Normal Bias Spring**     **Dead Weight Bias**     **Magnet Bias**

*Fig. 11. Different types of proportional and reverse bias forces*

To get the most efficiency out of an actuator wire, it is also possible to use another length of Flexinol to get opposite actuator movements, so restoration will be induced by the activation of each other.

## 3.4 Heating and Cooling Mechanisms

As shown in previous sections, the contraction of the actuator wire is due solely to heating and the relaxation, solely to cooling. Both contraction and relaxation are virtually instantaneous with the temperature of the wire. As a result, mechanical cycle speed is dependent on and directly related to temperature changes. Applying high currents for short periods of time can quickly heat the wire. It can be heated so fast in fact that the limiting factor is not the rate at which heating can occur but rather the stress created by such rapid movement. If the wire is made to contract too fast with a load, the inertia of the load can cause over stress to the wire. To perform high speed contractions, inertia must be held low and the current, applied in short high bursts.

While each device has quite different heat sinking and heating requirements, a simple rule of thumb test can be used to prevent overheating. Measuring the actual internal temperature of the wire across such short time periods is somewhat problematic. However, one can tell if the actuator wire is overheated simply by observing if the wire immediately begins to cool and relax when the current is shut off or not.

If it does not begin to relax and elongate under a small load promptly when the power is cut, then the wire has been needlessly overheated and could easily be damaged. Simple visual observation is all that is needed to design measured heating circuitry in prototypes.

All the conventional rules for electrical heating apply to the wire, except that its resistance goes down as it is heated through its transformation temperature and contracts. This is contrary to the general rule of increased resistance with increased temperature. Part of this drop in resistance is due to the shortened wire and part is due to the fact that the wire gets thicker as it shortens, roughly maintaining its same three-dimensional volume. It makes no difference to the wire whether alternating current, continuous current or pulse-width modulated current is used.

The simplest way to improve the speed of cooling is to use smaller diameter wire. The smaller the diameter the higher surface-to-mass ratio the wire has and the faster it can cool. Additional wire, even multiple strands in parallel, can be used in order to exert whatever force is needed. The next factor in improving the relaxation or cooling time is to use higher temperature wires. Those wires contract and relax at higher temperatures. Accordingly the temperature differential between the ambient or room temperature and the wire temperature is greater and correspondingly the wire will drop below the transition temperature faster in response to the rate of heat loss.

Other methods of improved cooling are to use forced air, heat sinks, increased stress –this raises the transition temperature and effectively makes the alloy into a higher transition temperature wire– and liquid coolants. Combination of these methods are also effective. Relaxation time can range from several minutes –what can be useful for delay switches but rarely for robotics– to fractions of milliseconds by effective and prosper heat sinking.

## 3.5 Wire Diameter

The complexity for heat dissipation when working with thicker wires make the lighter diameters particularly attractive for robotics projects. Actuators made with 0.006" wires and below can easily be controlled by basic microcontroller techniques and powered by simple transistor arrays such as the familiar Darlington transistor ULN2003 and ULN2803. To achieve greater pull from the thinner wires, more actuators can be used in parallel, or even mounted as an array an controlled together.

## 3.6 Electronic control circuit

A nitinol wire can be activated through low voltage, between 6 and 12 volts. The simplest circuit is shown in Fig. 12. It consists of a battery, a switch and a length of nitinol wire. It is very important to take into consideration that a continuous current through the nitinol wire will continue increasing its temperature thus degrading its properties.

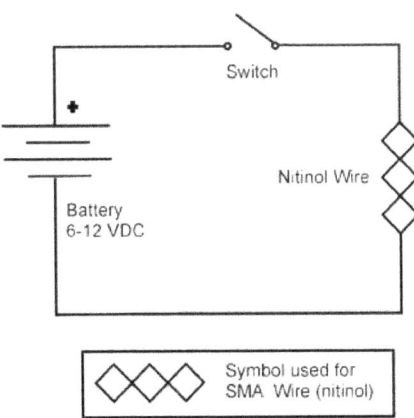

*Fig. 12. Basic circuit for nitinol testing*

Continuous current does not heat the wire homogeneously. To that, it is recommended to use a pulse-width modulation (PWM) circuit, as shown in Fig. 13. That circuit is the recommended to actuate nitinol wires for robotics, not just for security reasons but also because it is possible to control the percentage of extension desired for the wire by setting the pulse width.

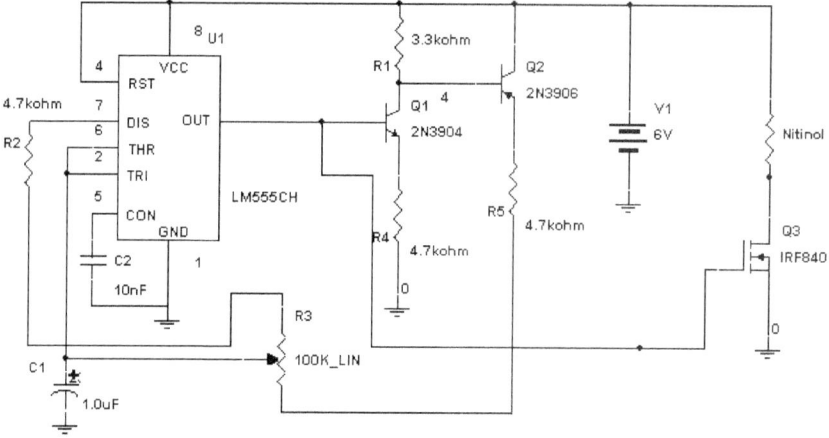

*Fig. 13. PWM circuit for nitinol actuation*

If the current through the nitinol wire is too high and produces too quick reaction for the load to move, a resistor can be put in series with it. Please take in consideration the maximum power of the resistor, as the passing current may be high.

# 4  APPLICATIONS OF SMA IN ROBOTICS

The use of SMAs as artificial muscles allows for more realistic bioinspired actuation presented in nature. SMA wires acting as muscle fibers can respond upon electrical signals, taking advantage of the large pull force and its excellent strength-weight ratio. Currently, the use of SMAs in biomimetic robotic systems can be found in ground, water, aerial and space robots, in many sizes, including those micro-robots or microstructures. In this chapter, the most representative bioinspired robots and structures that integrate SMAs as muscle-like actuation mechanisms are described.

## 4.1 Robots for Water

Biologically inspired robots that operate in water can be found in two categories. Firstly, robots that use SMAs for actuating appendices (fins), and secondly, robots that use SMAs to actuate the robot's body. In the latter, body actuation is used for undulatory motion (fish-like robots). Some animals can move by bending their body in such a way to produce a backward-propagating propulsive wave. The movement obtained by bending a continuous structure is much more natural than others where joints are presented.

### 4.1.1  Micro-Fish Robot

It has been proposed a micro-fish robot that uses a flexible biomimetic fin propeller with embedded SMA wires to mimic the musculature and flexible bending of squid fin. The propulsion consists of an active component –the biomimetic fin– and a passive component –the caudal fin. The biomimetic fin-based propulsion mechanism is an actuator that combines the

SMA wire and an elastic substrate.

This micro-fish robot introduces a new concept in the world of biomimetic robotics, due to its ability to swim noiseless. This means the robot avoids the use of any traditional components like gears, bearings and joints, only using the SMAs as actuators that produce the propulsion.

The robot is able to achieve a swimming speed of 112 mm/s when the SMA wires actuate at 2.1 Hz, contracting upon electrical heating, and a minimum turning radius of 136 mm, which makes this robot the fastest micro-fish robot compared to other prototypes that use IPMCs as actuators.

## 4.1.2    Small-Scale, Water-Jumping Robot

There has also been developed a water-jumping robot, that mimics the ability of water striders and fishing spiders of jumping on water surface. This biomimetic robot achieves a vertical jumping motion by pushing the water surface. The motion is triggered with a latch driven by an SMA actuator.

As a result of the research, quantification of several technical parameters made on the laboratory suggest that the physics of jumping in this robot is similar to that of the fishing spider. In terms of actuation, the SMA allows the robot to be extremely light (mass of 0.51 g), which it is essential to ensure the buoyancy on water. The maximum jumping height is 26 mm, 26% of the height reached when jumping on ground (53.1 mm). This prototype is the first concept of jumping robot that integrates SMAs within a structure with an overall mass of 1 g. The robot has to be fed with 2 W of power in order to generate a force of 1.35 mN.

## 4.1.3    Manta-Ray Fish Robot

There has been also built a manta-ray fish robot, actuated by SMA wires. Fig. 14 shows the prototype of the robot. Two pectoral fins arranged in triangular shape, made of latex with a thickness of 0.2 mm, form the fin surface.

*Fig. 14. Micro-biomimetic manta-ray fish robot*

This micro-manta ray is the first prototype that uses SMAs to generate ·thrust. This robot is capable to swim forward and turn. The pectoral fin sweep-back angle is 20°. A maximum swimming speed of 57 mm/s has been achieved, with a maximum amplitude for the motion of 40 mm. All the biomimetic fins are open-loop controlled.

## 4.1.4   Lamprey-Based Robot

A sea lamprey has been recently developed. The robot consists of a cylindrical, electronic bay, propelled by an undulatory body axis. SMA actuators generate propagating, flexion waves in five undulatory segments of a polyurethane strip. The lamprey robot, shown in Fig. 15, consists of a cylindrical hull that houses the electronics and battery pack. In this application, a neuronal network is used control the robot in real time. This neuronal network also drives control of the SMAs. The experiments have shown the system can reject disturbances thanks to the robustness of the nonlinear controller. Each SMA wire drains 1.5 A of electrical current when activated.

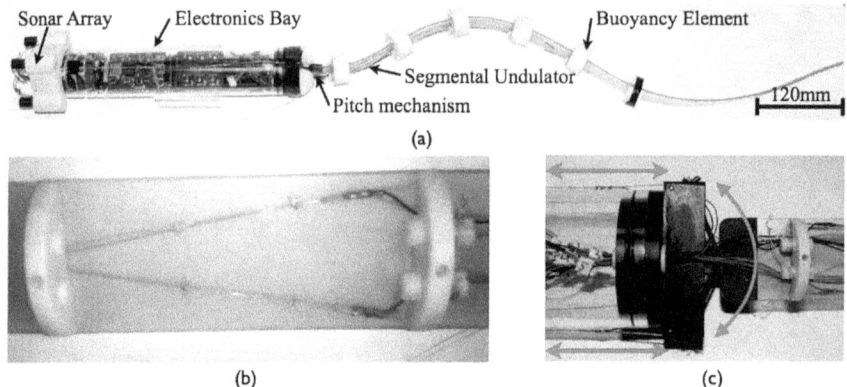

*Fig. 15. Lamprey robot (a), tail segment detail with nitinol actuator (b) and pitch mechanism (c)*

## 4.1.5   Biomimetic Robotic Jellyfish

The newest and more advanced aquatic robot that uses SMA actuators is a jellyfish robot. The hydrogen-powered robot, called "Robojelly", mimics the propulsion, morphology, kinematics and physical appearance of a real jellyfish. The bio-inspired actuators are made of flexible silicone, SMA wires and steel springs.

The development of Robojelly has introduced a systematic method for the design and fabrication of SMA-based actuators called BISMAC (Bio-Inspired Shape-Memory Alloy Composite). This method allows for bending the structure of the robot by means of SMA contraction. Thanks to the BISMAC SMA arrangement, this robot is capable to mimic the physics and swimming characteristics of a jellyfish in terms of its bell geometry, passive relaxation mechanism, neutral buoyancy, frequency of motion, and deformation-to-flap motion profiles. The structure can be bended by the SMAs actuators (deformation), and then a flap motion of the bell-segment structures provide the propulsion.

Robojelly is able to produce enough thrust to propel itself and operate over 5 cycles per second, which is comparable to the natural medusa at 4 cycles per second. The robot consumes an average of 16.74 W per 14th cycle of actuation. This robot confirms the benefit of using SMA actuators, combined with other materials, to create movement in aquatic, biomimetic robots.

### 4.1.6 Biomimetic Tuna

An SMA-actuated tail, inspired by a blue-fin tuna, is shown in Fig. 16. This fish-like robot has a length of 1 m and is able to swim in the biological carangiform type. The maximum tail beat frequency is 0.5 Hz, due to the limitations of SMA. Even at this low frequency, power requirements are significant. The minimum and maximum power consumptions are estimated in 292.8 W and 333.6 W, respectively.

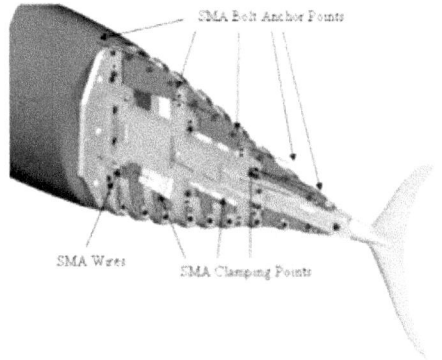

*Fig. 16. CAD representation of Suleman's tuna tail*

## 4.2 Aerial Robots

In aerial, bio-inspired robots most of SMA applications are appendices. Robots can be assessed in two main categories: insects and birds. Up to date, only a single flying-insect robot has been developed. This can be explained by the high flapping frequency needed, far from SMAs capabilities and also by the consequently high power requirement. For these reasons, insect-like flying robots mostly adopt piezo-electric actuators. In fact, the flying insect prototype described below uses SMA to fold and unfold the wings, and not for the primary flapping motion. Despite the speed of SMA actuators does not allow flapping wings, it can help in other kind of wing movement, such as morphing-wings.

## 4.2.1  Beetle Robot

A beetle-like insect robot, inspired by the Allomyrina Dichotomapresents, is presented. This robot features a morphing-wing airfoil capable of folding and unfolding the hind wing using SMA wires. A single, small size DC motor drives the flapping mechanism. Fig. 17 shows the prototype and the unfolding of the artificial flapping/morphing wing device.

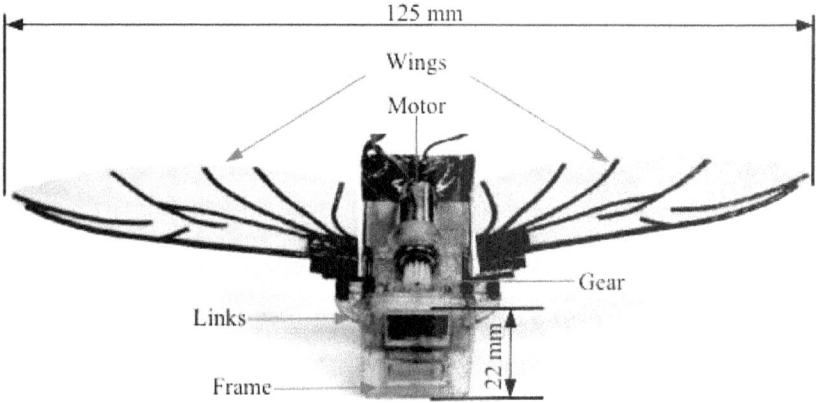

*Fig. 17. Prototype of the robotic beetle*

Folding speed of the robot wings is 1.7 times slower than their biological counterpart's. On average, wing unfolding is completed within about 3 s and folding, in about 4 s.

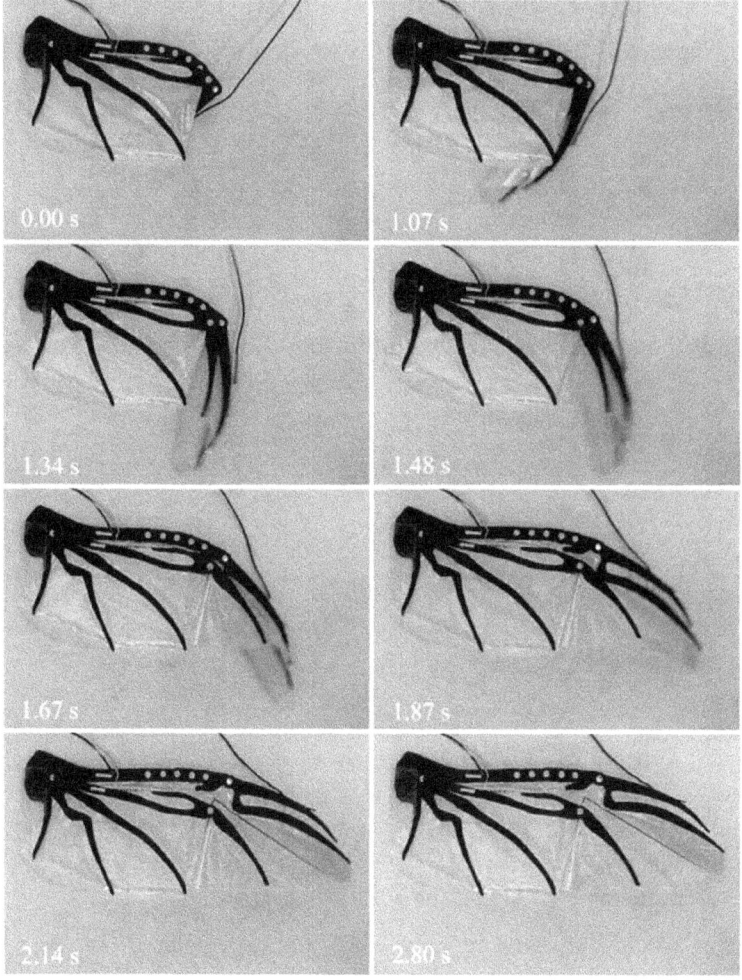

*Fig. 18. Detail of the unfolding process or the wings*

## 4.2.2 Bat Robot

BATMAV is a biologically inspired bat-like MAV (Micro-Aerial Vehicle) with flexible and foldable wings, capable of flapping flight. This robot has bat-inspired wings, with a large number of flexible joints, that allow mimicking the kinematics of a real bat. Fig. 19 details the overall robot structure and SMA muscle-like actuators main connections.

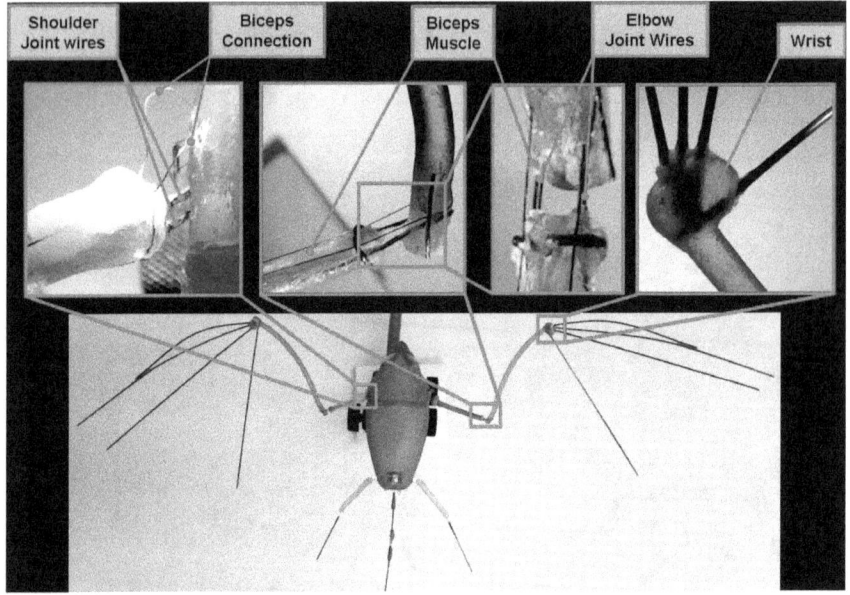

*Fig. 19. BATMAV with SMA wires as actuators and super-elastic joints*

BATMAV is the first robot that uses the SMA wires to play a dual role: first, as muscle-like actuators, providing the flapping and morphing wingbeat motions of the robot and, second, as super-elastic flexible hinges that join the wings bone structure. Most of the experiments have been carried out with a two-degree-of-freedom wing, capable of flapping at 3 Hz.

Despite the fact that the robot is able to achieve accurate bio-inspired trajectories, the results presented lack experimental evidence of aerodynamics measurements that demonstrate the real utility of the proposed design.

## 4.3 Ground Robots

Ground bioinspired robots have been divided into two categories: the ones that use actuated appendices (i.e. legged robots) and those that actuate the whole body, like crawling robots (snakes, worms, etc.).

### 4.3.1 SMA-Based Hexapod Robot

SMABOT is a hexapod biomimetic robot, with two SMA actuators that provide with a two degree-of-freedom movement to this robot. Each actuator produces 300 g of pull force. Fig. 20 shows the SMABOT IV, which incorporates a two-dimensional inertial navigation system for position control. Average speed when moving with its tripod gait is 30 cm/min and maximum power consumption is about 25 W, with a mass of 290 g.

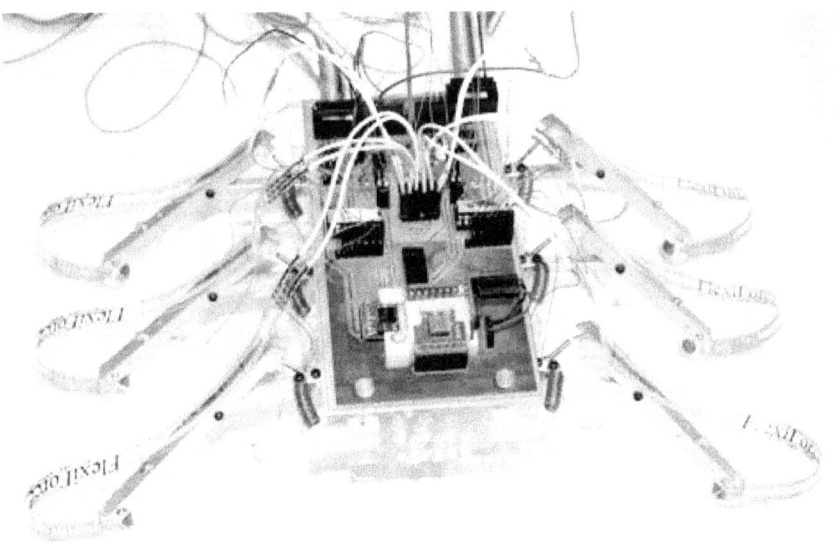

*Fig. 20. SMABOT IV, a SMA-based hexapod robot*

### 4.3.2 Crawling Robot Inspired by Ascotis Selenaria

Omegabot is a robot inspired by the inchworm Ascotis Selenaria. This robot is named after the omega ($\Omega$) shape of the crawling motion of the inchworm.

Fig. 21 shows Omegabot platform. Experimental results report the first step for an inchworm-like robot that can crawl on various terrains, where conventional robots cannot move. Omegabot uses a SMA coil actuator that requires a current of 200 mA for activation. The frequency of motion is about

1 Hz, limited by the response time of SMA wires. The inchworm robot is manually controlled by an IR remote operation, and it achieves a maximum linear velocity of 5 mm/s, although the robot travels a distance of 5 mm per stroke.

*Fig. 21. Omegabot, a biomimetic inchworm robot*

### 4.3.3   Another Earthworm-Like Micro-Robot

There is also a bio-mimetic micro earthworm-like robot with wireless control. The actuation mechanism consists on a SMA spring that contract and extend the earthworm muscle respectively. The proposed mechanism is simple but effective when traveling in narrow and rough environments, such as human digestive organs, bended long pipeline and so on. Also, this micro robot incorporates both control and power supply onboard.

The theoretical speed of this micro-robot is approximately 3.4 mm/cycle, with a total cycle time of 8 s (SMA contraction time is 2 s and recovery time is 6 s). The prototype can move at a speed of 10 mm/min for 8 minutes. The stroke per cycle is 2.0 mm.

# 4.4 Robotics Parts

In this section two works that do not address the development of a full robot, but rather appendices to be added to future robotic systems, are presented.

## 4.4.1 Flexible Pectoral Fin

Experimental research on pectoral fin structure has been performed recently. The design of the pectoral fin actuator is based on SMAs wires, composed by a couple of plates with opposite functions. Fig. 22 shows the biomimetic pectoral fin schematics.

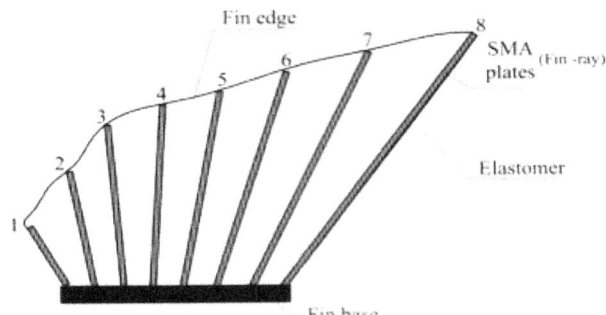

*Fig. 22. Biomimetic pectoral fin driven by 8 couples of SMA plates*

This pectoral fin was the first and remains being the only designed with just SMA wires. In fact, most research on fish-like robots is focused on studying propulsion (how to generate thrust), while maneuvers is largely unexplored.

## 4.4.2 Shape-Memory Alloy Actuated Biomimetic Hydrofoil

There has been developed and tested a biomimetic active hydrofoil, using Shape Memory Alloy (SMA) actuators. It consists of a six-segment demonstration foil with its control schemes.

# 4.5 iTuna: a Swimming Robotic Fish

iTuna is a swimming fish-like robot that, including its external appearance, imitates some key features of fish morphology.

This mechatronic concept takes inspiration from the arrangement of the red or slow-twitch muscles (see Fig. 23). In live fishes, such muscles are used for bending a flexible but nearly incompressible axis. Such axis is either composed of a visco-elastic beam (notochord) or a series of vertebrae connected through intervertebral discs. The main structure of iTuna robot is inspired by the former solution, and it is composed by a continuous, flexible backbone. This backbone is composed of 1 mm-thick polycarbonate, actuated by SMA muscles, as red muscles.

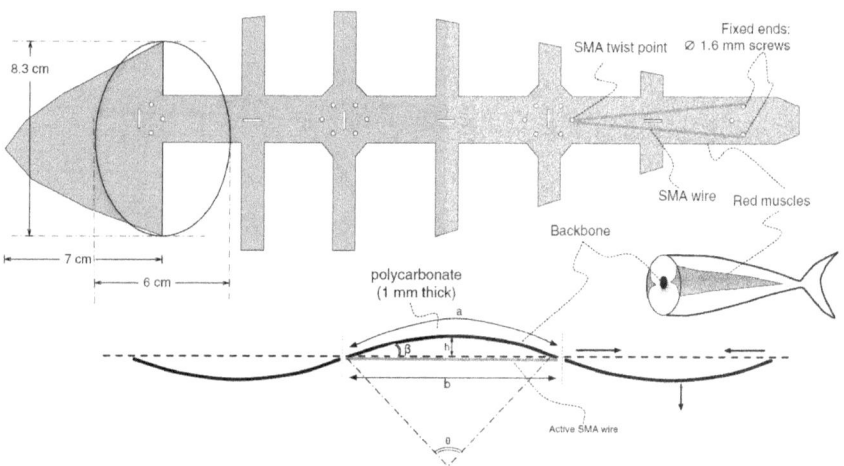

*Fig. 23. Structure of iTuna robot*

Six SMA-based actuators, with 1/3 the length of the body, are located in pairs, parallel to the body in a way to produce antagonistic movement on three body 8.5 cm-long segments. This antagonistic configuration of SMA wires has some advantages in terms of increasing the range of controllable actuation, since both directions of motion (contraction and elongation) can be actively controlled. Fig. 23 shows the location of the SMA wires within the skeleton structure of the prototype.

A V-shape wire configuration, where each artificial muscle is composed of a single V-shape SMA wire, twisted around the tension screw, allows to double the pull force without a significant increase in power consumption. Nitinol wires with 150 μm diameter have been used. These have a pull force of 230 g when fed with 250 mA at room temperature, and a nominal contraction time of 1 s.

Under nominal operation the SMAs can bend the body segments up to 28° (angle β in Fig. 23), even when SMA wires only approximately contract by 4% of their length. By increasing the input current and including a suitable control to handle overloaded operation of SMAs, contraction time of 0.5 s can be achieved, and strain can be fostered to 6%, corresponding to a bending of 36° (Fig. 24).

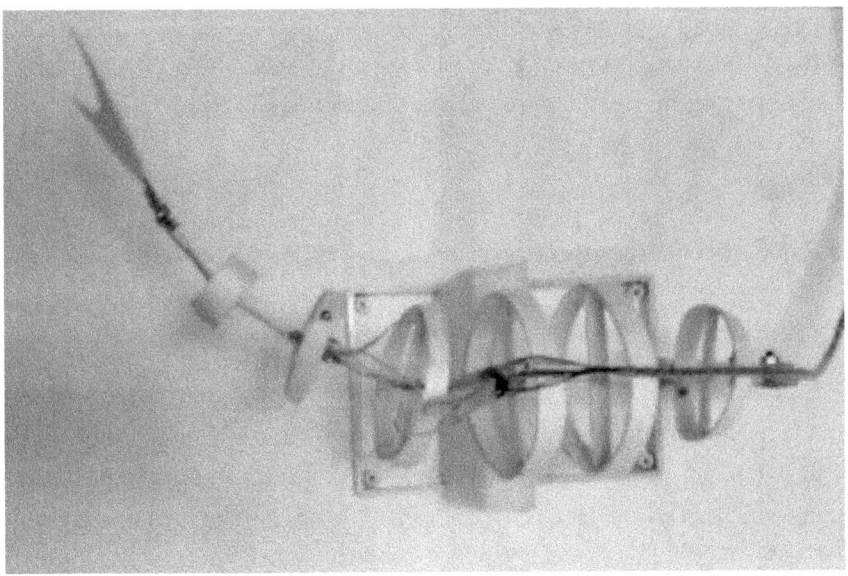

*Fig. 24. Increased bending when SMAs are overloaded*

### 4.5.1   SMA Control in iTuna

A low-level PID controller has been designed to address two main limitations of SMAs: slack in the fibers and limited actuation speed. Slack issues appear when SMA wires develop a two-way memory effect during operation. Limitation in actuation speed occurs due to the large switching time between cooling and heating phases. To address these problems, a pre-heating mechanism has been developed, which works in conjunction to the antagonistic arrangement. The pre-heating avoids temperature on both wires drop below 10% the maximum applied electrical current, preventing the inactive alloy from complete cooling. On the other hand, the antagonistic arrangement provides an external stress to the cooling wire, provided both by the elastic backbone and by the active antagonistic wire. Working with an already-warm wire allows for a faster stretch and slack issues are avoided.

Please note the PID controller is based on the experimental observation of the fact that the hysteresis on the electrical resistance curve is smaller than the hysteresis on the temperature curve.

Resistance measurements are used as a feedback signal for closed-loop control. The control developed allows overloading the SMA with up to 350 mA peak current (please note power signals are sinusoidal, hence overloading only lasts a brief period of time). Overloading has allowed for achieving a 1 Hz oscillation time (i.e. 0.5 s for each contraction and cooling times) and a 36° bending angle of each body segment.


## 4.5.2   Control Architecture

A key feature of SMAs is the possibility to develop closed-loop control systems without the need of external sensor hardware. The feedback signal is provided by the detection of inner electrical resistance, that allows an indirect measurement of temperature.

The main components are described in the following. A micro controller implements the PID algorithm. The PID controller receives the input reference position (set point) and the feedback of SMA voltage and current that allows calculating the heating current to drive the SMA actuator. The PID controller digital output is converted to a reference current in two steps. Firstly, it is converted into an analog signal using a 2-wire serial 8-bit DAC (Digital to Analog Converter) with rail-to-rail outputs. Then, a Voltage Controlled Current Source (VCCS) transforms the DC voltage into a constant current that feeds the SMA. This stage has a power consumption below 10 mA.

The DAC output ranges from 0 to 5 V with a resolution of 0.02 V. The measured voltage (VSMA) and current (ISMA) on the SMA fiber are fed back to the micro controller in order to close the control loop. The hardware used (PIC 16F690) has a 12-bit A/D converter with a resolution of 0.537 mV, considering the maximum voltage measured at the SMA VSMA = 0.55 V.

On the other hand, taking into consideration the maximum current through the wire (500 mA), SMA resistance variations about 1.074 m$\Omega$ can be measured. Therefore, since maximum variation in SMA length is 0.34 cm, and maximum variation of resistance is 1.6 $\Omega$, the theoretical position error of the system based on the SMA length is 0.067%, i.e. 0.12 mm.

## 4.6 BaTboT: a Biologically-Inspired Bat-Like Robot

BaTboT is a bio-inspired bat robot that uses SMAs as artificial muscles for powering the morphing motion of the wings. The morphing motion is related to the robot ability to modulate its wings, by contracting and extending the membrane in sync with the flapping motion. It is precisely this characteristic what makes biological bats more agile to maneuver than any other flying creature in the same Reynolds number range (103−104). In addition, biological studies have revealed that bats are able to maneuver thanks to the inertial changes produced by wings modulation. By mimicking this feature using an artificial device, BaTboT mainly presents a twofold challenge: first, the wings biomechanical design, and second, a proper control and actuation to module them.

Fig. 25 shows the design-flow process to evaluate key issues of SMA performance and its implications to the robot.

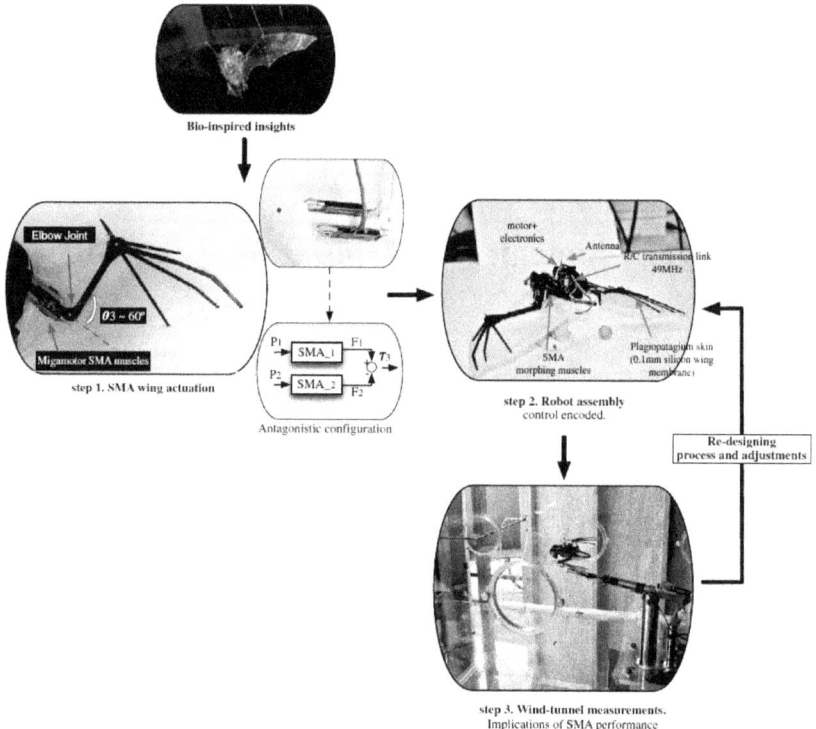

*Fig. 25. Flow process for SMA evaluation in BaTboT prototype*

## 4.6.1　Working Dynamics

The SMA actuators shown in Fig. 25 (step 1) are the commercial Migamotor Nano Muscle model RS-70-CE. Each nano muscle consists of several short strips of nitinol wire, with a diameter of 150 μm, attached to opposite ends of six metal strips, stacked in parallel.

Each SMA segment pulls the next strip about 0.67 mm relative to the previous strip, and the relative movements sum to make a stroke of 4 mm. As depicted, two Migamotor muscles have been arranged in an antagonistic configuration, working as artificial biceps and triceps, that provide the rotation motion of the wing elbow joint. The motion range of the joint is about 60°. The joint motion is achieved when each SMA actuator contracts upon heating, thus generating a pull force. As both actuators are connected to the joint in an antagonistic way, the pull force generates a joint torque.

Therefore, each actuator requires an input heating power to produce and output torque.

## 4.6.2   Actuation Speed and Fatigue

A PID controller is sufficient for allowing accurate and faster SMA actuation. The key issue for achieving an outstanding SMA performance does not necessarily depend on the control technique, but in complementary mechanisms that monitor the lower and upper limits of input power.

The improvement in accuracy and speed are due to two factors: firstly, by avoiding wire entanglement and, secondly, by ensuring the passive wire of the antagonistic configuration does not cool completely. Wire entanglement can be produced when the SMA wires are extended upon cooling. The passive SMA wire can develop a few millimeters of slack as it cools, which consequently affects the accuracy of the control. This slack-phenomenon is only presented in the antagonistic configuration, due to the two-way shape memory effect produced by each actuator.

To avoid the aforementioned issues, the anti-slack mechanism defines a minimum threshold of input heating power that ensures the inactive wire does not cool completely. The improvement in actuation speed is due to the fact that the already-warm SMA wire can begin to pull as soon as the heating current is raised, whereas a cold wire would first need to be raised to its operating temperature. It has been observed, from experimental results, that a suitable minimum threshold of input heating power is about 10% the power applied.

On the other hand, the anti-overload mechanism is in charge of ensuring the maximum input power does not increase above a top limit. This approach avoids overheating the SMAs in case the controller delivers a large amount of power to the wires. This upper limit can be found using a phenomenological model, or by performing real measurements of SMA temperature and stress on the wires.

Both anti-slack and anti-overload mechanisms are key for improving on SMA performance under a force control architecture. There are two advantages of using a force-control scheme: a high-bandwidth response and SMA fatigue avoidance. High-bandwidth response requires the use of force sensors, capable of providing the force feedback. It has also been demonstrated that, by using high-bandwidth force feedback, limiting cycles of SMA operation are eliminated.

Nevertheless, for some systems, the use of force sensors can be a hardware limitation. It has been demonstrated that both anti-slack and anti-overload mechanisms can be implemented in a position control scheme. The position feedback can be achieved by measuring the electrical resistance of the SMA wires, which is a linear function of the strain.

The key disadvantage of using a position scheme that forces the SMA to behave in overloaded operation mode relies on fatigue. As experimentally observed, overloaded operation mode could be maintained only for about five minutes of SMA continuous operation before decreasing performance to nominal mode. For the application at hand, overloaded mode implies an actuation frequency of 2.5 Hz, while nominal mode, an actuation frequency of 1.3 Hz. In this case, fatigue issues causes a decrease in actuation speed performance about 56%.

# 4.7 Humanoids

## 4.7.1   Robotic hand

While 3D printing has already proved itself in the field of mechanical hand prostheses with very basic or no functions, the field of bionics has proven more challenging. To be sure, there are a number of very promising ongoing projects out there, but all reach the same obstacle: how do you make a bionic cheap, functional and lightweight at the same time? It looks like, in 2015, a team of German engineers from Saarland University have come up with a bionic hand that relies on a bundle of smart wires and an electric charge instead of bulky motors and electronics.

Key are these bundles of very thin muscle-like fibers that have been made from nitinol wire. These wires are some of the best actuation mechanisms in the world for featuring the highest energy density out of all competitors, as seen before. This enables very powerful movements in miniscule spaces. Additionally, the fact that they come in the form of wires enables to use them as artificial muscles, or artificial tendons.

While conventional bionics are filled with motors that govern the opening and closing movements of a prosthetic, these wires return to their original shape after being deformed, meaning that the 'off-mode' can be an open hand, and a single electric charge is all that is needed to close it. When turned off, the wire efficiently returns to the open mode without using energy.

Each of the bundle has been designed to copy the structure of a human muscle. To do so, all the fine wires have been grouped into bundles, just like muscle fibers. Each bundle is capable of undergoing rapid contractions and releases, just like a human hand muscles are capable of. In this design, ach phalanx can be moved independently.

Again, like in the examples above, the shape of the wires themselves is controlled by a single chip, which functions on electrical resistance. This means that no external sensors are needed, as the material itself has sensory properties that govern movement very precisely. This is especially useful for picking up objects. This helps in always preserving a very lightweight structure. This is a big deal, because normally prostheses until now are very heavy.

Unfortunately, this prosthetic is still in its prototyping stage, but the developers team is very hopeful about the results and envision prosthetic limbs that function and feel like regular arms and hands. Especially the removal of bulky motors and pneumatics means that they can also be easily worn by everyone.

Considering organic-silicon synapsis, currently being under study, there is also the possibility of integrating these smart strands into a person's own neurological system.

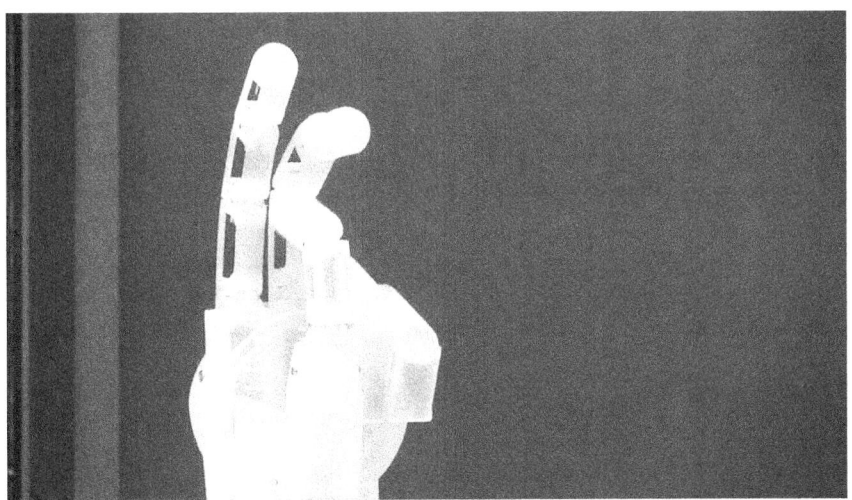

*Fig. 26. Robotic hand with muscle wire and 3D-printed structure*

## 4.8 Baby Humanoid Head

A biped mountable robotic baby head was developed using a combination of Biometal fiber and Flexinol shape memory alloy actuators (SMAs). SMAs were embedded in the skull and connected to the elastomeric skin at control points. An engineered architecture of the skull was fabricated, which incorporates all the SMA wires with 35 routine pulleys, two firewire complementary metal-oxide semiconductor cameras that serve as eyes, and a battery powered microcontroller base driving circuit with a total dimension of 140×90×110mm3140×90×110 mm3. The driving circuit was designed such that it can be easily integrated with a biped and allows programming in real-time. This 12DOF head was mounted on the body of a 21DOF miniature bipedal robot, resulting in a humanoid robot with a total of 33DOFs. Characterization results on the face and associated design issues are described, which provides a pathway for developing a humanlike facial anatomy using wire-based muscles. Numerical simulation based on SIMULINK was conducted to assess the performance of the prototypic robotic face, mainly focusing on the jaw movement. The nonlinear dynamics model along with governing equations for SMA actuators containing transcendental and switching functions was solved numerically and a generalized SIMULINK model was developed. Issues related to the integration of the robotic head with a biped are discussed using the kinematic model.

## 4.9 Applications in Soft Robotics

Soft robotics is a newly born field that is now starting showing first convincing results. In this framework, the coupling of soft and flexible materials with SMA actuators has been successfully implemented, especially in biomimetic and bioinspired robotics. Caterpillars, worms and octopus arms are the most cited and studied. Reproducing the capabilities of such animals requires, in fact, the presence of a soft body with active abilities of motion without the use of rigid structures.

Among several caterpillar-like robots, one of the most effective examples is represented by the GoQBot. It replicates a very fast self-propelled wheeling behavior, producing a ballistic rolling by changing its body configuration in 250 ms. The soft robot is composed of three functional components: a soft silicone body, molded in a straight and long geometry (bar-like), two independent SMA actuators fixed in parallel respect to the body, allowing its anterior and posterior flexion, and a pair of tail skids providing stability. The control system is based on a pulse-width modulation (PWM) driving, with high power stimulations, in order to achieve a very impulsive response.

Earthworms inspired peristaltic locomotion platforms and contracting robots, that exploit antagonistic arrangement and elastic properties of the soft materials they are made of. As the example in Fig. 27, a flexible braided mesh-tube structure is used as a support for SMA springs, arranged in antagonistic manner: circularly, to squeeze and elongate the structure, and longitudinally, to shorten it. Sequential activation of the spring groups leads to peristaltic motion that can be used as a locomotion system. 400 mA current with short pulse time is experimentally found to be the most effective way to activate the springs and the coordination of the segments is obtained by a purposely developed gait controller. Closed loop control is implemented with the measure of the radius changes, by means of Hall effect sensors.

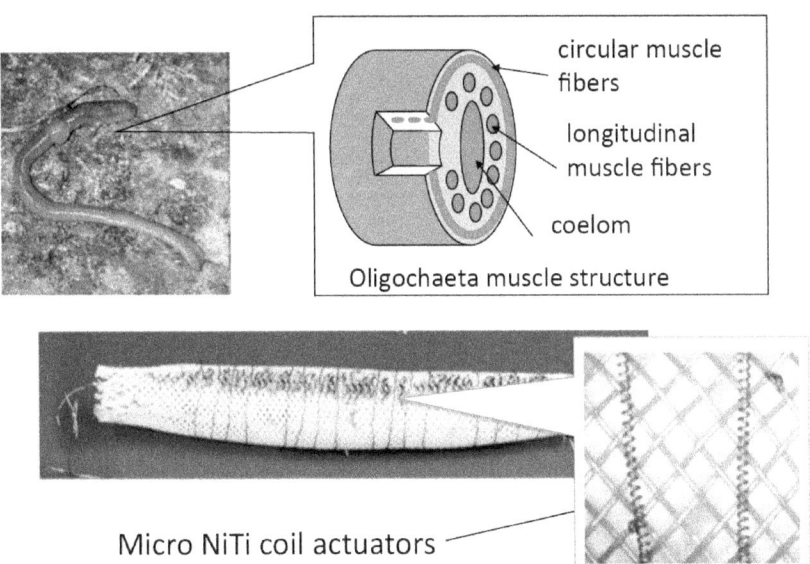

*Fig. 27. Mesh worm prototype with peristaltic locomotion*

A different approach has been also used to generate a forward locomotion: a TWE SMA spring located inside a cylindrical silicone bellow. The activation of the spring lets the structure shorten and the bellow delivers the bias force necessary to support the return phase of the alloy. The control system is open-loop and the right timing for cooling and heating has been experimentally set. The robot is completely autonomous and embeds battery for activation and wireless communication for on/off signals.

Another example of SMAs successfully used as actuation technology for soft robotics is represented by the development of an artificial octopus-like arm. The aim of the work was to develop an octopus-like arm with similar motion capabilities respect to the natural counterpart, by reproducing the fundamental characteristics that make it move so dexterously. The key factor which determines the performance of the octopus arm is the *muscular hydrostat*, a peculiar structure with muscles arranged on longitudinal and transverse planes, which can deform and vary it stiffness with contractions. Applying the principles of the muscular hydrostat led to the design of an artificial muscular hydrostat, that is based on a conical braided sheath that is used as the body of the arm and as a support for the SMA-based actuation system, as shown in Fig. 28.

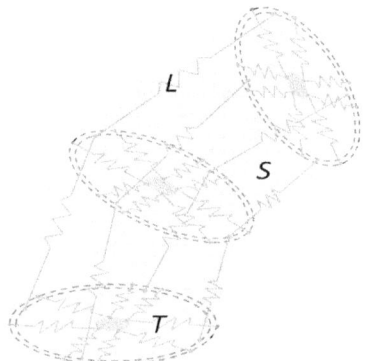

*Fig. 28. Schematic design of the artificial muscular hydrostats, with longitudinal (L) and transverse (T) SMA actuators*

Helicoidal springs have been chosen to meet the requirements of force and deformation. The transverse and longitudinal actuators act as antagonistic elements, since the former is able to squeeze the braided sheath, causing an elongation of the structure, and consequently of the longitudinal actuator too, while the latter is able to shorten the arm and to increase the diameter, restoring the transverse actuators. By manufacturing a long structure that counts on several sections like those depicted in Fig. 28, an entire octopus-like arm can be reproduced (Fig. 29).

*Fig. 29. Entire octopus-like arm*

The combination of the action of the transverse and longitudinal actuators allows the reproduction of the octopus basic movements, like elongation, shortening, bending in any direction –by activating asymmetrically the longitudinal springs– and stiffening –by contracting simultaneously both kind of actuators. An on/off activation is provided using PWM in an open-loop control. The robot has been designed to work in marine environment, so that an external skin is necessary to separate the external surrounding liquid and the internal controlled environment. Internally, the braided sheath and the actuators are immersed in distilled water, which increases the electrical power need but guarantees a very fast cooling phase.

Other than bioinspired and biomimetic robots, the use of SMA in soft robotics is exploited for implementing innovative locomotion strategies. Although rigid platforms like crawlers and legged robots are often used over rough terrain, an alternative method that exploits the deformability of the robot body is shown in Fig. 30. The robot is basically composed of a deformable circular shell and eight SMA springs radially arranged inside it. When the springs contract, the circle transforms into an ellipse, becoming instable, so that gravitational potential energy generates a moment around the point of contact between the body and the ground. This causes a body roll until it achieves a new stable state. The provided activation is on/off through a PWM and, with the appropriate voltage pattern, a continuous roll is achievable. The same approach has been adopted with some modification in the mechanical hardware of the circle and on the disposition of the actuating elements: eight flexible support structures are kept in circle by SMA sheets. In this case, the deformation of the structure is due to the activation of every laser-cut SMA sheet, that exerts a torque between the two flexible structures it connects. Once again, the activation method is on/off, but providing direct current and with some safety features to ensure avoiding overheating.

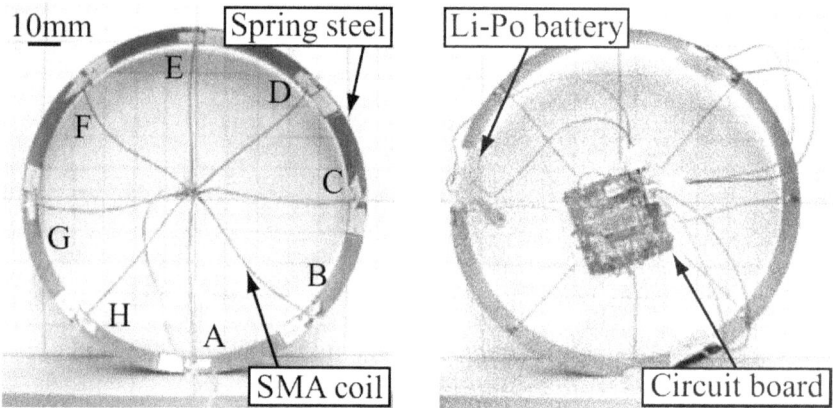

*Fig. 30. Rolling soft robot*

One last example is represented by a very recent and original work where SMA are used in a robotic origami, in Fig. 31: the kind of robots are able to fold into 3D shapes, starting from a nominally 2D sheet. A stretchable circuitry based on liquid-metal-filled channels in an elastomer substrate is integrated with a tiled origami module, actuated by a SMA actuator. In this case, the activation of the alloys is not directly achieved by Joule effect but with the use of external heaters, to avoid power losses in the circuitry path.

*Fig. 31. 32-tile self-folding sheet, capable of achieving two distinct shapes*

# 5   PRACTICAL EXAMPLES

In this chapter, some practical examples are provided for illustrating how nitinol muscle wires can be used in robotics applications. The purpose of this book is to provide global understanding about shape-memory alloys use in robotics, so there will be different level of details for the next examples. It is not intended to be a thorough guide for building nitinol devices, but a general information source to start a project where nitinol is involved.

## 5.1  Basic Gripper

This example is provided to notice the importance of the right design when nitinol is involved. The need for a bias force adds a dimension to robot design. Different elastic materials and different spring implementations will produce dramatically different results and some study of the field is necessary to develop sufficient understanding.

Creative problem solving will be needed in more complex situations. As a simpler case, consider the basic robotic gripper illustrated in Fig. 32. It uses Flexinol and a rubber band around a pivot. The rubber band functions as an extension spring keeping tension on the Flexinol and keeping the gripper in a "normally closed" state. The device only opens when power is applied to the wire.

The importance of this design resides in how this configuration allows the gripper to grasp objects of diferent sizes using the elastic power derived from the rubber band. If we had chosen the opposite design, were the Flexinol was responsible for pulling the gripper to the closed position, then it would need to go to a fairly precise location and could only grasp objects of a specific size, or else risk damage to the object or the Flexinol.

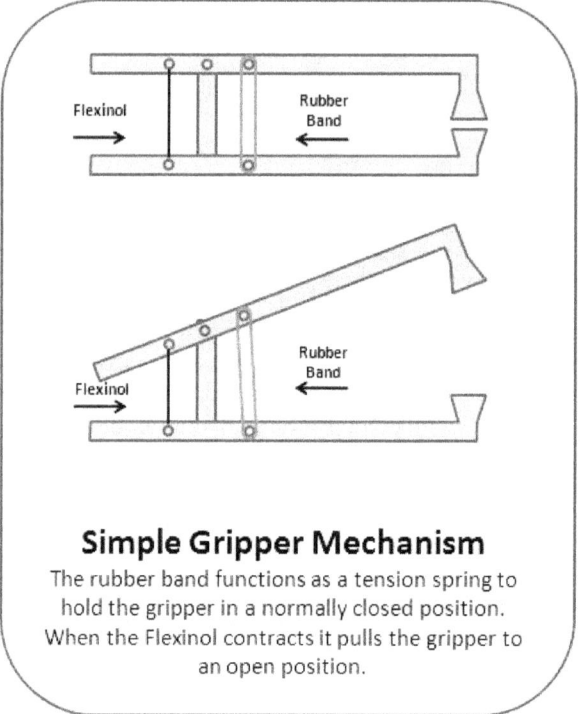

*Fig. 32. Basic gripper made with Flexinol*

Another relevant aspect that can be observed in this example is the use of leverage. The perhaps more familiar use of the lever and pivot is to allow a small force moving over a large distance to create a large force moving over a short distance. However, it is equally valid to use a lever to do the opposite and enable a large force moving over a short distance (Flexinol wire contraction) to create a small force moving over a large distance. This second application is exactly what happens in the operation of the gripper.

## 5.2 Microcontroller-Driven Flexinol

This case describes a microcontroller circuit to drive Flexinol muscle wires using a ULN2003A integrated circuit and a PIC 16F690. The method and circuit presented here provide a simple, low-cost solution with good flexibility well suited to the popular 0.004" and 0.005" Flexinol wires. It should hopefully be useful in a variety of robotics projects.

The PIC 16F690 used here as an example is a good all-purpose controller from Microchip Technology's midrange offerings. It is a 20-pin 8-bit processor with flash memory and a number of enticing peripheral features, including 12 analog channels, 3 timers, PWM, and I2C. It is the chip that ships with the "PICkit2 starter kit" and there is a good amount of sample source code around. Of course the basic interface described here can easily be adapted to any microcontroller platform such as Arduino, PICAXE, BASIC Stamp or MSP430 Value Line.

The ULN2003A is an array of 7 Darlington transistor pairs in a 16-pin DIP package. The IC includes not only the Darlington transistor pairs, but the necessary suppression diodes and current limiting resistors to create the full circuit. For many microcontroller applications, including this, it is possible to drive loads by connecting outputs from the controller directly to inputs of the ULN2003A without any additional components. Together with its other virtues, the ULN2003A is a tidy little chip. Each of its seven inputs on the left side of the chip connects to an output directly opposite on the right. In addition to the seven inputs and outputs, pin 8 is wired to ground, and pin 9 is optionally wired to positive voltage when sinking inductive loads. Pin outs for both the PIC 16F690 and ULN2003A are pictured in Fig. 33 and Fig. 34, respectively.

**PIC16F690 Pin Diagram (PDIP, SOIC, SSOP)**

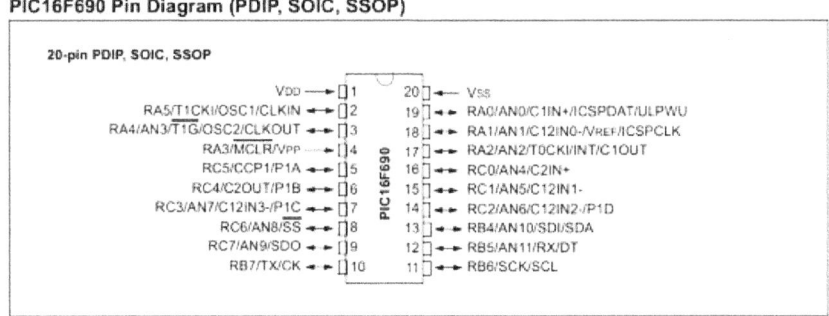

*Fig. 33. PIC16F690 pinout*

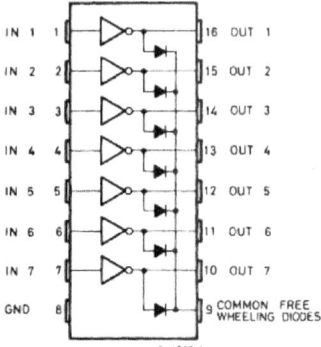

*Fig. 34. ULN2003A pinout*

Each of the ULN2003A outputs are rated for a maximum load of 500 mA while the whole array can sink as much as a full amp depending on configuration and ambient temperature. When controlling multiple outputs simultaneously you must use the graph in Fig. 35, reproduced from the datasheet, to determine the maximum current available per output.

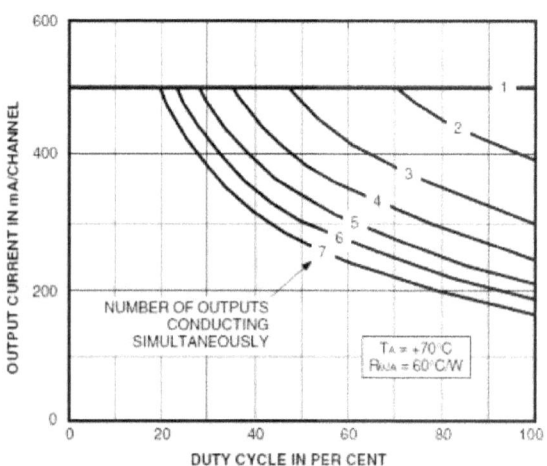

*Fig. 35. Output current for ULN2003A outputs*

The graph plots curves are based on duty cycle, but for this basic circuit we will assume a 100% duty cycle and just use the far right of the curves. Therefore one simultaneous output will have a maximum current of 500 mA, two outputs about 390 mA each, three outputs about 300 mA, etc. The 0.004" Flexinol needs about 200 mA to achieve a single second contraction. Using the graph we could determine that up to five outputs could be sunk simultaneously at 200 mA. However, it's best not to push quite that close to the maximum, so four outputs is a better working number. Remember that this restriction is for outputs conducting simultaneously. More than four outputs can be connected to 200 mA loads, provided that they do not need to fire at the same time.

A schematic illustrating the basic Flexinol driver circuit appears in Fig. 36. It is quite simple and hopefully, for the most part, self-explanatory. The schematic shows four connections for clarity, but an actual application could have more or fewer, and could be tied to any available outputs of the PIC.

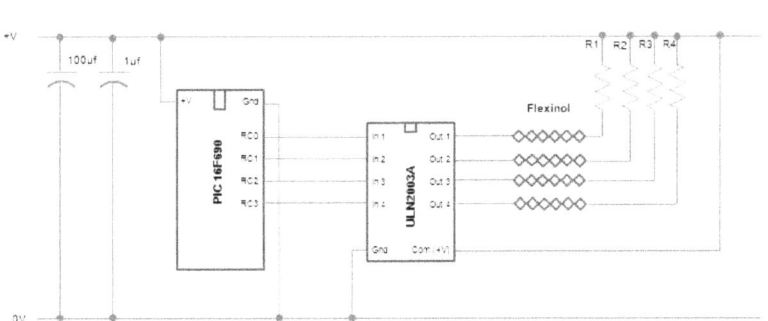

*Fig. 36. Four Flexinol wires control circuit*

The values for the resistors have to be calculated based on the supply voltage, the desired current, and the inherent electrical resistance of each length of Flexinol. When making this calculation, the voltage drop across the ULN2003A must be accounted for. The drop is not a constant but will range dependent upon collector current as can be seen in the graph in Fig. 37, which is reproduced from the datasheet. The graph shows both typical and maximum values for Vce.

## SATURATION VOLTAGE
## AS A FUNCTION OF COLLECTOR CURRENT

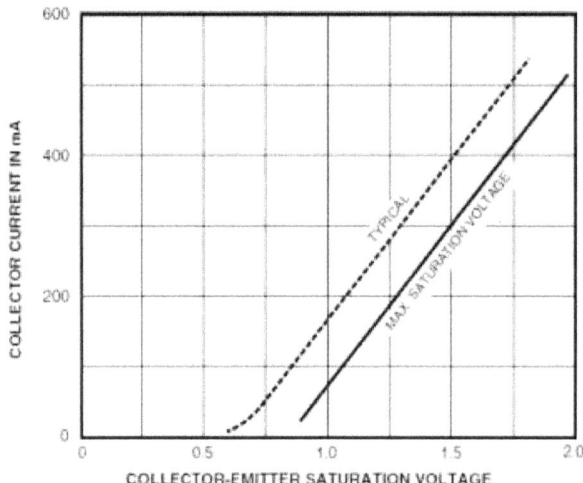

*Fig. 37. Saturation voltage of ULN2003A transistors*

As an example calculation, assume a 5 volt power supply and a two inch length of 0.004" Flexinol supplied with current of 200 mA. Using the saturation graph above, the ULN2003A will drop about 1.1 volts at 200 mA, so we can assume that roughly 3.9 volts is being delivered to the circuit. Plugging these values into Ohm's Law (R=E/I) we get 3.9 V /0.2 A = 19.5 $\Omega$ resistance required. The 0.004" Flexinol has a resistance of 3.2 $\Omega$ per inch (see Fig. X), which is 6.4 $\Omega$ in our two-inch example. Subtracting 6.4 from the 19.5 total required leaves approximately 13.1 $\Omega$ to be made up in the circuit with resistors.

The sample code for the PIC below is elementary and purely illustrative, cycling through the outputs 0 to 3 on PORT C, which are shown connected in Fig. 36. The program uses delay loops and holds each pin high for about one second before advancing to the next state.

```
;-----------------------------------------------------------------
-
; flexi-drive.asm
;
; sample code in Microchip Technology PIC Assembly Language
; to accompany a circuit that controls Flexinol actuator wire
;
; individual pins RC0-RC3 are brought high for intervals of approximately
; one second and the cycle repeats endlessly
;-----------------------------------------------------------------
```

```
;-------------------------------------------------------------------------------
;
;  initialization - load the standard include file and set the configuration bits
;-------------------------------------------------------------------------------

list p=16F690
#include
__config (_INTRC_OSC_NOCLKOUT & _WDT_OFF & _PWRTE_OFF & _MCLRE_OFF & _CP_OFF
                 & _BOR_OFF & _IESO_OFF & _FCMEN_OFF)

;-------------------------------------------------------------------------------
;
;  variables
;-------------------------------------------------------------------------------

UDATA
  CounterA res 1                    ; these variables are all used as counters
  CounterB res 1                    ; in the delay subroutine
  CounterC res 1

;-------------------------------------------------------------------------------
;
;  Code
;-------------------------------------------------------------------------------

CODE 0x00

Start:
    banksel   TRISC
    clrf      TRISC               ; make PORTC all output
    banksel   PORTC

PinCycle:
    movlw     b'00000001'         ; put the bit pattern for the outputs into W
    movwf     PORTC               ; load the bit pattern into PORTC
    call      OneSecondDelay      ; go to the delay subroutine
    movlw     b'00000010'         ; repeat for the next pattern...
    movwf     PORTC
    call      OneSecondDelay
    movlw     b'00000100'
    movwf     PORTC
    call      OneSecondDelay
    movlw     b'00001000'
    movwf     PORTC
    call      OneSecondDelay
    goto      PinCycle


OneSecondDelay:                   ; the delay routine is based on a sequence
    movlw     0x06                ; of nested loops which have been calculated to
    movwf     CounterC            ; take approximately one second to complete
    movlw     0x18
    movwf     CounterB
    movlw     0xA7
    movwf     CounterA
OSDLoop:
    decfsz    CounterA,1
    goto      OSDLoop
    decfsz    CounterB,1
    goto      OSDLoop
    decfsz    CounterC,1
    goto      OSDLoop
    return

end
```

There are a few easy modifications that could be made to this circuit. If more outputs are needed, additional ULN2003As can be added or, if you need just one more, the ULN2803A provides eight Darlington transistors in an 18-pin package.

If you are considering substituting a ULN2003 or ULN2003L, first read the datasheets carefully as these chips have lower total power dissipation ratings. To sink higher currents than described here, outputs can be configured in parallel (doubled-up). However, the per-chip maximums will still apply. For projects with significantly higher current requirements, consider using MOSFETs or even relays to manage the load.

## 5.3 Precision Flexinol Position Control Using Arduino

As nitinol contracts when conducting electricity, an approach to precision control of this effect based on controlling the voltage in the Flexinol circuit is presented in this example. In addition, taking advantage of the fact that the resistance of Flexinol drops predictably as it contracts, the mechanism described here uses the wire itself as a sensor in a feedback control loop. Some advantages of eliminating the need for a separate sensor are reduced parts count and reduced mechanical complexity.

This design is built around an Arduino Uno. To drive the Flexinol, it makes use of a homebrew digital to analog converter (DAC) that can be assembled from common parts. To calculate the resistance, Arduino native analog to digital (ADC) peripheral is used to measure the voltage drop across Flexinol. The control code for the circuit first self-calibrates, and then steps through a set of targets attempting to hold positions across the range from fully relaxed to fully contracted. Throughout the run, data is sent to the Arduino serial monitor for observation and study.

Achieving precise, low-jitter position control of a Flexinol mechanism turns out to be surprisingly difficult. Although Flexinol contraction is usually driven by electricity, the shape memory effect that causes the contraction is actually a function of temperature. In an aptly named phenomenon called resistive heating, electric current makes the wire get hot and it contracts. All by itself, controlling resistive heating in a thin length of wire is elusive, involving variables such as ambient temperature, air flow, altitude and even humidity. Beyond those environmental basics however, controlling Flexinol requires a reasonably thorough understanding how the wire will actually respond to changes in temperature.

In the on-off model of Flexinol control used in previous examples, we typically give the wire more than enough electric current to heat it and fully transition to the contracted austenite phase. We then remove electricity entirely to allow the wire to cool and transition to the relaxed martensite phase. Fortunately for the effort at hand, Flexinol does not simply transform at a particular temperature, but rather the effect is incremental across a rather wide temperature range. The full temperature span varies by the formulation but is typically between 40 and 70 °C.

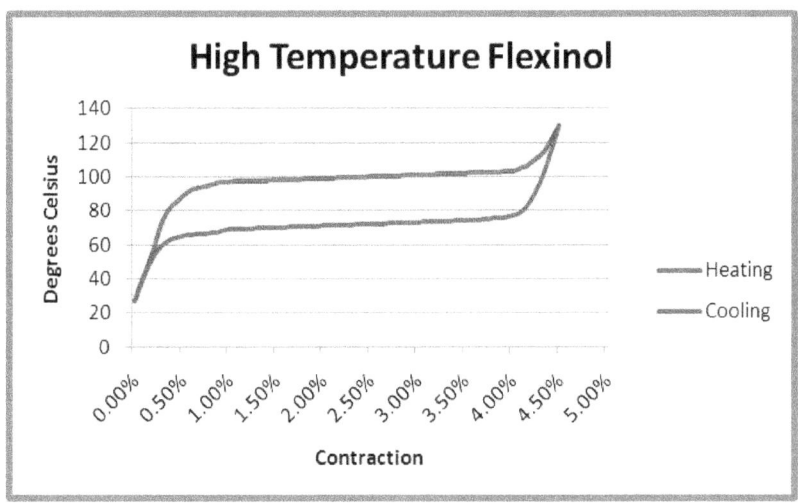

*Fig. 38. Flexinol contraction-temperature dependence*

Adapted from Flexinol specification document, the chart in Fig. 38 shows temperatures in °C plotted against the percentage of contraction for high temp (HT) Flexinol. You can immediately observe the temperature response of Flexinol is nonlinear. It begins gradually, accelerates to the point where most of the useful physical transformation takes place over about a 10 degree change in temperature and then tapers off. A second observation is that there is a significant difference in the transformation temperatures when Flexinol is heating to the austenite phase compared to when it is cooling to the martensite phase. When heating, the mid-point between fully contracted and fully relaxed is at about 100 degrees; when cooling, the mid-point is about 70 degrees. This 30 degree gap is known as the transformation temperature hysteresis, as seen before.

Also to be considered is the fact that, when the current being supplied to Flexinol is changed, the temperature does not change immediately, it needs time to either warm or cool. Therefore, when current is changed Flexinol will not move immediately to the new position, but will continue contracting or relaxing for a certain period. If we continue to change the current while the Flexinol is still settling, then we overshoot the target and the mechanism ultimately swings back and forth across the target rather than holding steady. Overshoot is further complicated by transformation hysteresis and heat dissipation issues. Typical approaches to feedback control where the current is proportionately increased or decreased until the desired position is reached will not work very well here.

In the solution presented here, transformation temperature hysteresis is accounted for by calibrating such that for a given position there is both a predetermined warming voltage value and a predetermined cooling voltage value. If Flexinol is too relaxed to hold the target position then the warming value is fed to the DAC. If it is too contracted for the position, then the cooling value is fed to the DAC.

The warming and cooling values are determined in a calibration process each time the program is run. The first step in the calibration is to establish a baseline by measuring the resistance of Flexinol when it is fully relaxed and fully contracted. In the second step, voltage to Flexinol is steadily increased and then steadily decreased, all the while calculating the resistance of the wire. When certain target resistance values are reached, the DAC values are stored in a lookup table. In making the calibrations, the current is changed at a moderate pace, a compromise between giving Flexinol time to settle at the new position while still moving quickly enough to be useful. Although this approach does not completely eliminate overshoot, by sampling relatively continuously at high resolution and switching between warming and cooling current as needed, jitter in the mechanism can be reduced to trivial levels.

## 5.3.1   Circuit Setup

As shown in Fig. 39, the mechanical part of the experimental setup consists of a ~5.9" length of Flexinol attached to the short end of a lever. When Flexinol contracts, the arm rocks upward; when it relaxes gravity acts as the bias force and stretches it back into shape. In this example, HT Flexinol 0.005" diameter wire has been used. According to the specifications, this wire has a resistance of 1.9 $\Omega$/inch. That means a current of 320 mA will cause it to fully contract in about a second and it will relax in about 1.4 seconds once the current is removed.

*Fig 39. Illustration mechanical structure for the experiment*

On the electronics side, current to Flexinol is controlled by varying voltage. Because we'll be measuring resistance in the circuit, we need a nice clean voltage. A resistor ladder DAC (frequently written R-2R) will be used to control the voltage. An R-2R DAC is a clever construct that enables you to build up a DAC of theoretically any resolution simply by adding *rungs* to the ladder –8 rungs equals an 8-bit DAC. Binary values are fed into the ladder by toggling eight digital pins from the Arduino and a proportionately corresponding voltage is available at the output. The ladder is a resistor network built up of two values, one the double of the other. Here, 10K and 22K resistors have been used. The other components of the DAC portion of the circuit are an LM358 operational amplifier and the PN2222 transistor. They are both setup as voltage followers in order to get enough current gain to supply Flexinol. Arduino can be powered directly through the USB programming cable and the DAC can be fed with a separate 6 V power supply. Finally, the circuit shown in Fig. 40 feeds a voltage divider made up of a fixed value resistor (2.7 Ω) and Flexinol, which acts as a variable resistor.

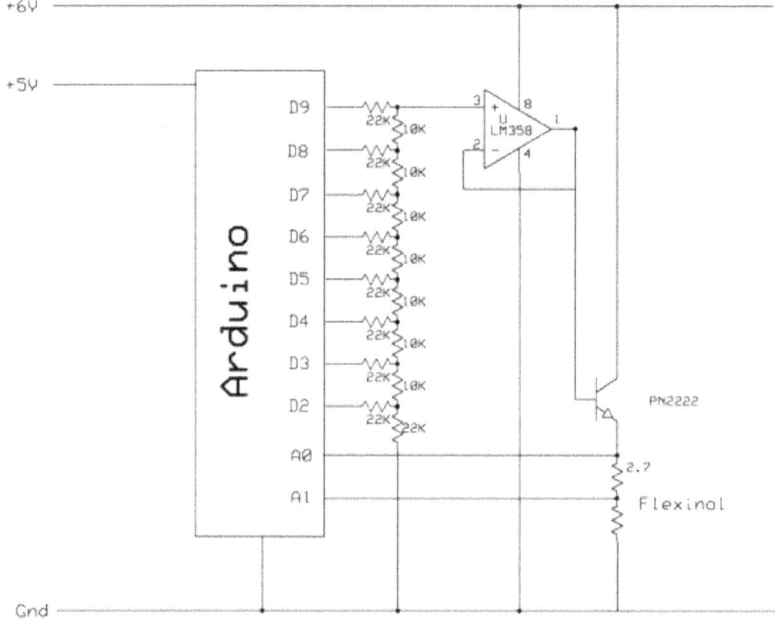

*Fig. 40. Electronics for the experiment*

## 5.3.2   Calculating the Resistance

In a voltage divider, when the value of the first resistor is known, the value of the second resistor can be calculated based on the voltage drop across the circuit using the following formula:

$$R_2 = \frac{R_1 \cdot V_{Out}}{V_{In} - V_{Out}}$$

This calculation is pretty easy to code on Arduino. The first step is to take readings using `analogRead` at A0 and A1. Those readings can then be converted to voltages by multiplying the values by 0.004883, which is equivalent to 5/1024 or reference voltage/ADC resolution.

Although the calculation itself is simple enough, it is here that we run into the limitations of both the test circuitry and the 10-bit ADC of the Arduino. According to the specifications, the difference in resistance between austenite and martensite phase should be about 20%. Experimentally you will get other values. In this particular circuit the actual useful range will be closer to 15% or about 1.5 $\Omega$, for this length of wire. Therefore in order to produce useful results, measurements need to be accurate to within a tenth of an ohm, or better yet several hundredths. With this setup we just barely get there.

## 5.3.3   The Results

Although not perfect, the results of the experiment are quite good. According to data culled from test runs, the setup is able to reliably hold Flexinol to a variance of <0.1 $\Omega$ from the target value. Observing the mechanism while testing, there is occasional jitter when moving to a new position, but overall the arm is held solidly with no discernable movement. As it stands, the technique could be applied in numerous projects where incremental positioning is needed, but extreme precision is not required. There are many opportunities for improvement, some of which are simply a matter of substituting higher quality components and building permanent rather than prototype circuitry. It seems likely that continued experimentation and refinement of the approach in both hardware and software will enable greater precision.

## 5.3.4   The Code

Although this experiment is only for practice in a laboratory, the code here is reasonably well commented to help make some of the notions described in this book and get the ability to develop more complex projects.

```
/*****************************************************************************
This sketch is part of an experiment in precise control of Flexinol (also known
as Muscle Wire). The premise of this method is that because the resistance of
Flexinol decreases as it contracts, the resistance at any given moment can be
used to determine the current state of contraction.

Connected to a circuit and mechanism described in this example, an Arduino is
used to deliver a controlled voltage to a length of Flexinol while continuously
measuring the resistance. The sketch spans two phases. In the first phase the
circuit is calibrated and a set of voltage/resistance values is built. In the
second phase a method of using these values to control the position of the
Flexinol mechanism is tested.

Platform: Arduino Uno
Environment: Arduino 1.0
```

```
**********************************************************************/

#include <math.h>

const float rl = 2.7;  // Value of R1 in the Flexinol voltage divider
const byte msbPin = 9; // The highest numbered pin of the 8 bit output to the DAC

float vIn;             // The measured voltage going into the Flexinol voltage
divider
float vOut;            // The measured voltage between R1 and the Flexinol
float rFlexinol;       // The calculated resistance of the Flexinol

byte dac;              // The byte value to write to the DAC
byte thisPin;          // Used in the routine to set the individual bits of the
DAC
byte thisBit;          // Used in the routine to set the individual bits of the
DAC

float avgActual;       // Used in testing to track the average resistance
readings
float delta;   // Used in testing to track difference between target & actual
values

int index;                   // Used to reference array variables
float resistanceValues[11];  // Target resistance values
int dacValuesWarming[11];    // Calibrated DAC values when Flexinol is warming
int dacValuesCooling[11];    // Calibrated DAC values when Flexinol is cooling


/*****************************************************************************
Functional subroutines:

bytewrite:  takes an 8-bit value and sends each bit to an individual Arduino pin
            to drive the DAC

calcResist: takes ADC readings at A0 and A1 and uses the results to calculate the
            current resistance of the Flexinol
*****************************************************************************/
void setup() {
  Serial.begin(9600);
  pinMode (2,  OUTPUT);
  pinMode (3,  OUTPUT);
  pinMode (4,  OUTPUT);
  pinMode (5,  OUTPUT);
  pinMode (6,  OUTPUT);
  pinMode (7,  OUTPUT);
  pinMode (8,  OUTPUT);
  pinMode (9,  OUTPUT);
}

void loop(){
  calibrate();
  test();
}

/*****************************************************************************
Calibration:

This subroutine first measures the highest and lowest resistance readings across
the range of potential voltages. It then calculates a set of resistance values at
regular intervals. Finally DAC values are cycled through continuously testing for
the best value to achieve each of the calculated resistances.
*****************************************************************************/

void calibrate(){
  Serial.println ("Working");
  Serial.println (" ");
  resistanceValues[0] = 7; //Seed high&low resist.values with values well outside
  resistanceValues[10] = 13; // the expected range
  dac = 0;
  byteWrite();
  delay(10000);
```

```
/******************************************************************************
Find the high and low resistance values and a corresponding DAC value
******************************************************************************/

  for (dac = 100 ; dac <= 210; dac++)    // Range determined experimentally
  {
    byteWrite();
    delay (500);
    calcResist();
    if (rFlexinol > resistanceValues[0])
    {
      resistanceValues[0] = rFlexinol;
      dacValuesWarming[0] = dac;
    }
    if (rFlexinol < resistanceValues[10])
    {
      resistanceValues[10] = rFlexinol;
      dacValuesWarming[10] = dac;
    }
  }

/******************************************************************************
Intermediate resistance values (Flexinol positions) are calculated based on the
high and low.

The results are stored in an array and output to the serial terminal
******************************************************************************/

  for (index = 0; index <= 10; index++)
  {
    resistanceValues[index] = ((resistanceValues[10] - resistanceValues[0]) * ind
ex * .1) + resistanceValues[0];
    Serial.print (index * 10, DEC);
    Serial.print ("% Contracted Resistance = ");
    Serial.println (resistanceValues[index], DEC);
  }

  Serial.println (" ");

/******************************************************************************
Remove power and allow the Flexinol to cool
******************************************************************************/

  dac = 0;
  byteWrite();
  delay (3000);

/******************************************************************************
Set values 1-9 in the warming values table to maximum to test against
******************************************************************************/

  for (index = 1; index <= 9; index++)
  {
    dacValuesWarming[index] = 255;
  }


/******************************************************************************
Slowly increase voltage to the circuit - the first time a target resistance value
is recorded place the corresponding DAC value in the warming index
******************************************************************************/

  dac = dacValuesWarming[0];
  do
  {
    byteWrite();
    delay(500);
    calcResist();

    for (index = 1; index <= 9; index++)
    {
      if (rFlexinol <= resistanceValues[index] && dacValuesWarming[index] == 255)
      {
        dacValuesWarming[index] = dac - 1;
```

```
      }
    }
    ++dac;
  }
  while (dacValuesWarming[9] == 255);    // repeat until all values mapped
/*******************************************************************************
Send the results of the warming calibration to the serial terminal
*******************************************************************************/
  for (index = 0; index <= 10; index++)
  {
    Serial.print (index * 10, DEC);
    Serial.print ("% Contracted: DAC Value Warming = ");
    Serial.println (dacValuesWarming[index], DEC);
  }

  Serial.println (" ");

/*******************************************************************************
The cooling calibration simply reverses the above warming calibration, steadily
reducing the voltage and recording the DAC value the first time a target value
is reached

Note one difference: the cooling index checks to be sure that the cooling value
is in fact less than the warming value, and if not keeps cycling
*******************************************************************************/
  for (index = 10; index >= 0; index--)
  {
    dacValuesCooling[index] = 0;
  }

  do
  {
    byteWrite();
    delay(500);
    calcResist();

    for (index = 0; index <= 10; index++)
    {
      if (rFlexinol >= resistanceValues[index] && dacValuesCooling[index] == 0 &&
dac + 1 < dacValuesWarming[index])
      {
        dacValuesCooling[index] = dac + 1;
      }
    }
    --dac;
  }
  while (dacValuesCooling[0] == 0 && dac > 100);

  for (index = 0; index <= 10; index++)
  {
    Serial.print (index * 10, DEC);
    Serial.print ("% Contracted: DAC Value Cooling = ");
    Serial.println (dacValuesCooling[index], DEC);
  }

}


/*******************************************************************************
Test

The test routine is more straightforward than it may appear at first.  The
circuit cycles through the target resistance values and attempts to hold the
position. Resistance is calculated.  If the Flexinol is too relaxed, the warming
value is output to the DAC. If the Flexinol is too contracted then the cooling
value is used.

The avgActual and delta variables and associated calculations are used to track a
verage values and deviations from the target which are output to the terminal
along with the other data.
*******************************************************************************/
```

```
void test(){
  for (index = 1; index <= 9; index ++)
  {
    Serial.println (" ");
    Serial.print ("Testing ");
    Serial.print (index * 10);
    Serial.println("%");


    dac = dacValuesWarming[index];
    for (int outerLoop = 0; outerLoop <= 4; outerLoop ++)
    {
      delta = 0;
      avgActual = 0;
      for (int innerLoop = 0; innerLoop <= 49; innerLoop ++)
      {
        byteWrite();
        delay (200);
        calcResist();
        avgActual = avgActual + rFlexinol;

        if (rFlexinol < resistanceValues[index])
        {
          dac = dacValuesCooling[index];
        }
        else {
          dac = dacValuesWarming[index];
        }
        delta = delta + abs(resistanceValues[index] - rFlexinol);
      }

      delta = delta /50;
      avgActual = avgActual /50;

      Serial.print ("Target: ");
      Serial.print (resistanceValues[index], DEC);
      Serial.print ("   Average Actual: ");
      Serial.print (avgActual, DEC);
      Serial.print ("   Delta: ");
      Serial.print (avgActual - resistanceValues[index], DEC);
      Serial.print ("   Average Deviation: ");
      Serial.println (delta, DEC);
    }
  }
}


/************************************************************************
byteWrite takes the value in dac and outputs it to the 8-bit resistor ladder
wired to Arduino pins 2-9.
************************************************************************/

void byteWrite(){
  for (byte j = 0; j <= 7; j++){
    thisPin = msbPin - j;
    thisBit = bitRead(dac, 7 - j);
    digitalWrite (thisPin, thisBit);
  }
}


/************************************************************************
calcResist takes ADC readings at A0 and A1, converts them to voltages and uses
the results to calculate the resistance of the Flexinol — the constant value of
0.004883 used here is equal to the reference voltage of 5 divided by the ADC
resolution of 1024
************************************************************************/

void calcResist(){
  vIn = analogRead (A0); //take a reading to reset ADC after switching pins
  delay (10);
  vIn = analogRead (A0);
  vIn = vIn * 0.004883;
```

```
vOut = analogRead (A1);
delay (10);
vOut = analogRead (A1);
vOut = vOut * 0.004883;

rFlexinol = (r1 * vOut) / (vIn - vOut);
}
```

# 5.4 Antagonist Actuator for Accurate Positioning System

In this example, a special two-wire nitinol actuator is presented, to avoid the limitation of accuracy produced by the hysteresis existing in the single-wire actuators, which degrades system performance. In order to mitigate the effects caused by the hysteretic nonlinearities in SMA actuators, the development of control approaches for the positioning system with the SMA actuator is a challenging task. Focusing on this challenge, choosing a hysteresis model to describe the hysteresis phenomenon in SMA actuators is a primary step for controller design.

SMA actuators can induce a large force or displacements when temperature is changed. The displacement-temperature relation existing in SMA actuators has strong hysteresis nonlinearities, which exhibits a typical saturation curve. So far, numerous hysteresis models have been used to express the hysteresis phenomenon in SMA actuators but the most accepted ones are general operator-based models, because they present the advantage of reflecting the thermometrical hysteretic behavior of SMA actuators precisely, and can be easily combined with common control approaches to improve the performance of positioning systems.

These control approaches usually address two main tasks. The first is to compensate the effects caused by hysteresis nonlinearities in the actuators, such as establishing the inverse hysteresis model or treating the actuator as a time-delay system. The other is to ensure the stability of the close-loop control system.

Focusing on the proposed actuator, a general scheme is shown in Fig. 41, where an antagonist-type nitinol micro-actuator is presented. Antagonist actuators consist of two nitinol wires complementing each other. When a wire is heated, it contracts and induces tension in the other wire, and the pulley is displaced. When the other wire is heated, it recovers the strain and moves the pulley back in the opposite direction. Thus, the orientation of the pulley is determined by heating and cooling the two wires, alternatively. The purpose of this device is to set the pulley in a precise orientation angle.

When the nitinol actuator is heated, the force generated can be decomposed in two parts: stiffness –which is linear, depending only on physical characteristics of the wire– and contraction force –which is generated by a change in temperature and is a nonlinear function, presenting hysteresis and saturation as well.

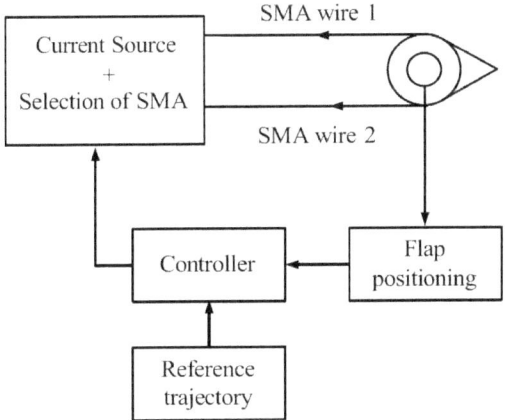

*Fig. 41. Antagonist, positioning system SMA actuator schema*

When performing a experiment with a close-loop controller and 23 cm wires, with 0.5 mm diameter –which can be deformed up to 10 mm–, the system can reach the desired position in less than 0.5 s and keep the steady-state error below 1%.

# REFERENCES

K. Byungkyu, G. L. Moon, P. L. Young, K. Yongin, L. Geunho, "An earthworm-like micro robot using shape memory alloy actuator", Sensors and Actuators A, 125, 2006, pp. 429-437.

M. Cianchetti, M. Follador, B. Mazzolai, P. Darío, C. Laschi, "Design and development of a soft robotic octopus arm exploiting embodied intelligence", ICRA'12, St. Paul, Minnesota, May 14-18 2012, USA, pp. 5271-5276.

W. Coral, C. Rossi, J. Colorado, D. Lemus, A. Barrientos, "Smart Actuation and Sensing Systems – Recent Advances and Future Challenges", InTech, 2012, pp. 53-82.

Y. Feng, C. A. Rabbath, H. Hong, M. Al Janaideh, C. Su, "Robust control for shape memory alloy micro-actuators based flap positioning system", Proceedings of 2010 American Control Conference, pp. 4181-4186, 2010.

N. Lechevin, C. A. Rabbath, "Quasipassivity-based robust nonlinear control synthesis for flap positioning using shape memory alloy microactuators". Proceedings of 2005 American Control Conference, pp. 3019-3024, 2005.

H. J. Lee, J. J. Lee, "Time delay control of a shape memory alloy actuator", Journal of Intelligent Material Systems and Structures, Vol. 13, pp. 227-239, 2004.

H. T. Lin, G. G. Leisk, B. Trimmer, "GoQBot: a caterpillar-inspired soft-bodied rolling robot", Bioinspiration and Biomimetics, vol. 6, 2011, 026007.

S. Majima, K. Kodama, T. Hasegawa, "Modeling of shape memory actuator and tracking control system with the model", IEEE Transactions on Control Systems Technology, Vol. 9, No. 1, pp. 54-59, 2001.

Y. Matsumoto, H. Nakanishi, S. Hirai, "Rolling locomotion of a deformable soft robot with built-in power source" CLAWAR'08, Coimbra, September 8-10, 2008, Portugal, pp. 365–372.

M. Moallem, V. A. Tabrizi, "Tracking control of an antagonistic shape memory alloy actuator pair", IEEE Transactions on Control Systems Technology, Vol. 17, No. 1, pp. 184-190, 2009.

J. K. Paik, R. K. Kramer, R. J. Wood, "Stretchable circuits and sensors for robotic origami", Proceedings of 2011 IEEE/RSJ International Conference on Intelligent Robots and Systems, pp. 414-420, 2011.

S. Seok, C. D. Onal, R. Wood, D. Rus, S. Kim, "Peristaltic locomotion with antagonistic actuators in soft robotics", ICRA'10, Anchorage, May 3-8 2010, Alaska, pp. 1228-1233.

S. M. Sutta, F. H. Ghorbel, "Differential hysteresis modeling of a shape memory alloy wire actuator", IEEE/ASME Transactions on Mechatronics, Vol. 10, No. 2, pp. 189-197, 2005.

J. Tadesse, D. Hong, S. Priya, "Twelve Degree of Freedom Baby Humanoid Head Using Shape Memory Alloy Actuators", Journal of Mechanisms and Robotics, Vol. 3, pp. 011008/1-011008/18, 2011.

E. Torres-Jara, K. Gilpin, J. Karges, R. J. Wood, D. Rus, "Compliant modular shape memory alloy actuators", IEEE Robotics and Automation Magazine, Vol. 17, Issue 4, 2010, pp. 78-87.

D. Trivedi, C. Rahn, W. Kier, I. Walker, "Soft robotics: biological inspiration, state of the art, and future research", Applied Bionics and Biomechanics, vol. 5, 2008, pp. 99-117.

BENJAMIN C. LAWSON

www.ingramcontent.com/pod-product-compliance
Lightning Source LLC
Chambersburg PA
CBHW060405190526
45169CB00002B/766